A CLOUD OVER BHOPAL

A CLOUD OVER BHOPAL

CAUSES, CONSEQUENCES, AND CONSTRUCTIVE SOLUTIONS

by
Alfred de Grazia

Metron Publications

ISBN: 978-1-60377-089-7
Library of Congress Catalog Number: 2014907481
Copyright © Alfred de Grazia, 1985, 2014
Second Edition
All Rights Reserved
METRON PUBLICATIONS
PRINCETON, NJ 08540-1213

To the memory of our dear friend,
Dr RASHMI MAYUR.

The First Edition of this book was published by the
Kalos Foundation for the India-America Committee for
the Bhopal Victims. Metron Publications, P.O.Box 1213,
Princeton, NJ 08542.

*The author wishes to express his thanks for the counsel given him by
the members and friends of the India-America Committee for Bhopal
Victims, especially to Dr. RASHMI MAYUR, Director, Urban
Development Institute, collaborator on this and other projects of
Kalos, and to Dr Kumudini Mayur; to Kevan Cleary, Tami
Aisenson, Ward Morehouse and Clarence Dias. Most important, my
wife, Anne-Marie de Grazia, collaborated with me throughout the
difficult months of work in the field and in the study.*

Publisher's Note of April 1985

Because of a wider expressed interest, this report, prepared for the use of the India-America Committee for Bhopal Victims, is being released to the public. Its publication may benefit those who need to know where matters stand and what issues now to pursue. The facts of the book are not final nor its judgements absolute. At this point in time, less than four months after the accident, many facts about the behavior of individuals, the deaths and damages, and the organization of responsibility have yet to come to light. The book is not written as a legal brief and ranges beyond the bounds of a court case. It presents hearsay evidence and many opinions and hypotheses waiting to be proven or disproved. Moreover, judicial processes will probably occur that have coercive means of eliciting documents and ascertaining facts. To avoid rendering judgements out of court and on the basis of the still incomplete evidence, the author has avoided, as far as possible, the use of proper names in the text. Despite these self-imposed limitations, the author stands firmly by his general interpretation and his proposals for compensation and reform.

As the gas cloud began to spread over the City of Bhopal in India shortly after midnight on December 3, 1984, equally tragic events were befalling humanity elsewhere in the world. To the West, a million Ethiopians were starving to death in the middle of civil wars. To the East of India, Vietnamese and Cambodian armies were slaughtering many thousands. To the Southeast, a violent ethnic conflict was upsetting the island republic of Sri Lanka threatening the lives and fortunes of thousands of persons. To the Northwest, two bloody wars were downing their victims by the thousands and four nations -- Afghanistan, The Soviet Union, Iran and Iraq -- were involved. We do not speak of troubles along the vast Northwest regions of the Indian border. Only over the great ocean to the South of India did peace reign, and uneasily at that.

Within India itself, the month preceding the Bhopal tragedy had witnessed the crazed killing of thousands of innocent Sikh Indians in the aftermath of the assassination of Prime Minister Indira Gandhi, the number of deaths being on the order of that visited upon Bhopal by the killing cloud of gas. A pall of psychic depression already hung upon the City. Nor should one overlook, in seeking to view the Bhopal case in perspective, the toll that pesticides, such as were being made at Bhopal, were taking around the world, an annual total of 10,000 fatalities and 375,000 poisonings in the Third World alone, according to the latest estimates; nor that a gas explosion a few days earlier had killed five hundred people in Mexico City, again poor people of the neighborhood.

These events are mentioned to fit the events at Bhopal into their place in a world society that cannot govern itself and take care of its people. But here we are charged to discover what happened at Bhopal. In that city, there occurred an immense and dramatic tragedy whose lessons are both local and worldwide. As we move out from the Center of India drawing upon these lessons, we can see the tragedy merging with the great stream of world tragedies that must be controlled all together, and the sooner the better, by a world power operating under a single benevolent and beneficent code of law and conduct.

I apologize to the victims for not describing fully their agonies and sorrows in this book. If I did, I could not possibly say all else that I need to say, which I believe to be in their interest and which is itself abbreviated. I realize also that the dying, the pain, the sorrow and the testimony are not yet ended.

Alfred de Grazia
Bombay, India
4th April, 1985.

Contents

CHAPTER I

In the Coolness of Night

The City of Bhopal is situated nicely upon the banks of two artificial lakes, created many centuries ago. It stands amidst the farms of a verdant plateau. It is the capital of the largest and center-most State of the Republic of India, Madhya Pradesh, whose population of some fifty-five millions contains a disproportionately large number of Moslems (there being 500 mosques in Bhopal alone), and also some ten million people still organized tribally. Per capita income in the State came to 1217 rupees (about $100.00) in 1981, one of the lowest averages in India.

The City must now consist of a million persons, or at least did so before the tragedy. Its industries are few, the factory of the Union Carbide Company of India, formulating pesticides, being outstanding for size and modernity. It holds a Medical School and a Technical College. Communications with the outer world are by train and bus, supplemented by several airplane flights a day that pause at the City's decent little airport before proceeding South or North. Automotive traffic on the highways to and from the city is slight. Still, enough traffic converges upon the several thoroughfares of the City to congest them.

Despite a number of elegantly constructed mosques and old houses, most of them approaching a desperation of disrepair, the total aspect of the City is unlikely to impress a stranger favorably. Actually there are few strangers, except from the villages around, and those who come to do business with the State Government, or to seek medical care at the free hospitals. (There is one hospital bed in the State for every four thousand inhabitants.)

On the more pleasant streets of the City the upper civil service, executives, proprietors and professional classes dwell; most people live in shabby but not unclean areas, and one wishes that he might go through town with a magical paint brush, because every wall, every interior, seems to want a coat of paint.

Unfortunately, when it came time for a sudden magic to strike Bhopal, it was not a bright paint, but instead an evil magic, a cloud of poison gas, disgorging from the best-painted, most modern part of the City, set apart from the rest by a tight, high wire fence, the Union Carbide complex, sixty acres of high technology complex on the northern outskirts of the City.

Is it superstition or some hidden rule that from the best-appearing the worst evil strikes? Was it not from Germany, cleanest and most advanced of countries, that there sprang suddenly the Nazi evil that almost destroyed Western civilization? Was it not from the immense factory of modern culture, the United States, where cities and farms are alike industrialized, that there appeared in the skies and upon the roads of Vietnam those splendid fleets of shiny aircraft and armored vehicles dealing out destruction to poor villages? For its own part, there is the high Soviet technology replicating the Vietnamese war in Afghanistan.

It is not superstition, but solid common sense, to pose the question of Bhopal in universal form : How can man use his most modern and ingenious developments in ways that will not turn upon his fellows and destroy them?

Every Indian city that can boast of progress must confess to the slums that come with progress. To create a new factory of the highest levels of design and technology employing one thousand workers in a fully modern setting is to create a shanty-town of 30,000 people. It is practically a sociological law, one which, however, has been given only cursory attention by those who try to build modernity by emplacing isolated scraps of it upon a traditional culture.

The phenomenon is not Indian, but worldwide. The phenomenon is also historic. As industry came to England, to France, Germany, old Russia, old America, it brought with it the same social movement. Myriads of people abandoned their villages to cluster around the new gods of industry so as to live a life that seemed a little better than the old one.

So the case of Bhopal is not unique, but typical, historically and today. One new "real" worker brings a family of half a dozen people, maybe more. These people attract more people. Some are relatives. Some are from nearby villages: they, too, hear the call of the city. There will be a little business to do because there is a little more money to circulate. Shops will be needed, and let us not be snobbish about what a shop must be like. A shop is a few vegetables; a shop is a man who fixes sandals with a hammer, pincers, a few nails and thongs; a shop is a person whose family goes out to find firewood, breaks it into little pieces, and sells it. In the end, a single well-employed worker will be supporting a family and auxiliary workers, and in part their families. We discover finally a ratio of one to twenty, and what began as a factory providing 1000 new job ends up supporting a large-sized town at the least, for we are not counting all the hopeful ones who move in and somehow add themselves to the already existing city economy, offering even cheaper services and labor. This ratio of "real" jobs is about one-fourth of the American ratio.

In the end, what it costs to keep one criminal in a New York State jail for a year, $40,000 (never mind the equal costs of putting him there in the first place) can keep 250 people going in India: civilized, decent, gentle, clean, and hardworking people who bear no grudge against the world. For such are the shantytown dwellers of Bhopal -- as elsewhere in India. And such are those who perhaps to the number of 60,000 (counting the rest of the victims as of higher socio-economic status) suffered most from the lowering and passing cloud of poison erupting out of high technology.

The cloud, called appropriately by some Indian newspapers "the killer-cloud", emerged in full hissing fury close to 1:00am from a venting tower, after passing through an apparatus designed to render harmless the poisonous gas of methyl-isocyanate (MIC). The gas ascended the vent pipe in a long-drawn-out explosion lasting for nearly two hours. It was initially propelled by the extremely high pressure of the tank that had held it in liquid form and emerged from the pipe into the atmosphere. Then, directed by the wind, it streamed out, not losing its internal turbulence until expanded and cooled. MIG has a density twice that of water, yet the cloud, both wind-propelled and self-propelled, could spread out far and wide while training its vapors along the ground.

The air temperature was in the mid-fifties Fahrenheit, a cool night for Bhopal. A fairly stiff breeze was blowing from the Northwest, from the countryside down upon the eastern sectors of the City. Both conditions -- the temperature and the wind -- were misfortunes: the chill air forced the hot and heavy poisoned moisture of the release to carry along close to the ground, preventing it from rising and dissipating.

The wind blew the gas through the most densely settled sectors of the city. About twenty-five square miles of territory were covered by lethal vapors during prolonged venting. Attempts at mapping the course of the cloud have produced differing configurations. A United Press International map, reproduced by the *New York Times*, pictured an overly well-defined course proceeding over adjoining shanty-towns, the railroad station and down through the southeastern part of the City. A map of B. K. Sharma of *India Today* presented a much more widely diffused pattern, which represented the many stories of poisoning in the downtown areas, and at the Straw Products Factory where many of the workers succumbed to gas, and at the hospitals and in the well-to-do areas. Figure 1 here follows the extensive view. No one questions that places most heavily affected were the congested slum colonies such as J. P. Nagar, Kazi Camp, Chola Kenchi, and Railroad Colony.

Even in the most tightly packed areas, the cloud did not behave uniformly. The variables that determined its fatality included the age and physical condition of the victims, one's sleeping posture, the position of one's mat or bed in relation to the open air, the varying use of cloths and water upon being struck by symptoms, and finally what must have been eddies, whorls, currents and pockets in the overall wave of poison.

The people knew right away the source of the poisonous air, although it was incredible and shocking. Thousands had fled their homes a few months before upon the occasion of a small discharge of gas and an associated rumor of disaster. Now they choked and screamed at one another to rise and flee, aiding each other when they could, the choking and gagging leading the fully blinded. Some stepped out of their huts at the first whiffs, strangling, and where too blinded to turn back in, were swept up in the gathering human torrent and often never saw their families, neighbors and friends again. Some fled in a fright that respected no one until they awakened as from a dream miles away. Havoc, chaos, madness in the mass: such words could be used for once literally.

No one ran toward the source of the cloud although to run against the wind would have been the rational action to take (as many who did not need to run said knowingly afterwards). Of course, they would have run up against the wire fence of the factory. To run crossways would imply that they would know the dimensions of the cloud; but so far as they knew the cloud might have been infinitely large. Wherever they turned they were met by a haze that worsened the choking, blindness and retching. Death and coma came as a final release from an excruciating agony, no matter whether of minutes or hours. The merely injured would continue to suffer hours, days and weeks of torture.

To the dead left behind were joined those dying along the line of rout. The crowd grew to be enormous and moved rapidly. In three quarters of an hour, its original surviving members had rushed five kilometers. This we know by figuring from the time of the gas release to the time when the Director of Medical Services, hastening to discover the trouble, was met by the onrush of people and had his car turned around and boarded by a score of victims, several of whom had expired but were mounted by others on the hood and top anyway. All obstacles were overrun; every cart, bicycle, and car was pressed into service. The incapacitated were sometimes trampled in the dark tumult. The dead, the screaming wounded, were everywhere one turned. Although the crowd hardly needed to be exhorted, a police van could be heard creeping ahead with its loudspeaker blaring: "Run for your lives! Poisonous gas is coming!"

Cattle, pigs, goats and dogs were exterminated in the path of the cloud. Later on their carcasses marked the contours of the cloud upon the ground.

The workers in the factory saw the venting of gas from its first occurrence and they could run against its flow. They numbered about seventy-five persons and certainly did not constitute a trained and disciplined force that would venture forth in their vehicles (which did exist) and protective gear (which also existed in limited quantity, the oxygen masks being of twenty-minutes duration) to give first aid (of what kind?) to the people crying out for help (in the dark inaccessible corners unreachable save by stretcher -- what stretcher?). Emergency help from the factory was nil.

The whitish greenish gas thus extruded was sufficient to have killed the million people of Bhopal had they been equally exposed and had the gas been sprayed uniformly in 180 degrees of arc. It is believed by toxicologists and it has been made into a rule in India and the United States that one in fifty million parts of methyl-isocyanate is enough to cause harm to a human exposed to it in an eight-hour work day. Twenty parts per million will send a person into agonies within five minutes. MIC is so reactive that experts Arthur Palotta and E. J. Bergin suspect it to be potentially a mutagen, teratogen, and carcinogen.

A tank of MIC thus becomes a kind of neutron bomb capable of fusion and explosion simply by adding water to it: people are destroyed and property is preserved. It is a wonder that terrorists have not targeted MIC installations, but perhaps untrained employees in charge of unsafe systems   can   also   do   the   job.

CHAPTER II

To Cope with a Killing Cloud

Late on the day of the disaster a figure of 600 deaths was being circulated. Each day for several days thereafter the number augmented by hundreds, and a statistical bifurcation manifested itself between conservative figures, which sought to stay under 2000, and radical figures, which began to soar over five thousand and achieved twenty thousand and even fifty three thousand, this last being a figure that was related to me by a respectable lawyer who claimed to have gotten it from a government geologist from Madhya Pradesh as well as other people who "ought to know".

Not without reason or support, I have settled upon a figure of 3000 dead, 10,000 seriously disabled, 20,000 significantly disabled, and 180,000 affected to minor degrees. Later on we shall have occasion to look at these figures more closely. Just now they are cited as a preliminary to describing how the institutions of the community responded to a disaster of such proportions.

The Union Carbide company was quite unready for the emergency. It could render no aid to people. For all the good they did, the thousand employees and the Indian and world network of 100,000 employees and hundreds of offices and factories and outlets might as well have been on holiday. This slight exaggeration is aimed at stressing the purely local nature of the disaster and the total response that was levied upon an unprepared and poorly equipped community, and should be qualified to mention the individuals who later on pitched in to help the community; it also implies the confused nature of the little medical and humanitarian aid provided, the perhaps deliberately misleading responses to police inquiries as to what was going on at the plant, and the aborted siren blast. The siren was first turned on full to give the public alarm, then quickly muted to the level of an internal alarm, and only raised to the public level after the gas cloud had been discharged. This last matter should be investigated thoroughly. The action was in keeping with the company policy of reassurance to the authorities and public; it could well have been ordered, and in any event was a decision by a company employee.

The only excuse for what would appear to be criminal negligence here is stupidity and panic. One needs to discover the mechanism of the siren to see whether it consisted of two sirens, or one that could be raised or lowered in volume of sound. If two sirens, an excuse is practically inconceivable. If the latter, a single siren that could be adjusted, then the operator might reason falsely, thus: 'Loud is for the public; soft is for the plant; so first I will turn it on loud to warn the public and then I will turn it on soft to warn the plant. There is no way of warning both at the same time.' Anyhow, the men in the MIC unit area warned the other employees by siren, loudspeaker, and voice to flee and they all did so and in the right direction.

The top man, the Plant Manager, was alerted around 1:45 A.M. by a city magistrate who presumably heard some news from the police, who in turn had been alerted by the actual seepage of the vapors into the city police control room and by a roving inspector who reported in around 1:00 A.M. that some disturbance was occurring at the Union Carbide plant. The Manager reached the site quickly, passing through a mist of MIC, which he opened his car window to sniff, and which caused him to tear and cough. By the time he reached the plant, the air there was breathable. He said that the plant telephones were not working. If so, it was because they had been abandoned, because the police had put through calls before his arrival and they had been told that the problem was under control. The national union leader of Union Carbide workers claims that the Plant Manager and Chemical Plant Manager called All-India Radio to alert and instruct the people, but the Radio station refused to alarm the public without permission of the Central Office.

Until the truth became all too obvious, plant officials claimed that they had stopped the leak in the fatal Tank 610, even adding that the feat had been accomplished in under an hour. Much later, it became apparent that the "leak" had not been plugged, but that all the MIC of Tank 610 had been exhausted, as cleanly as if by plan. Only a little vapor greeted the Indian Central Bureau of Investigation when it sought samples from it days later. The Plant Manager, with refreshing frankness, declined to call the accident a leak, as the New York Times and many of the press and public figures termed it, insisting that it was an 'uncontrolled emission.' In effect he was calling it a constrained, slow-motion explosion.

If there had occurred no other way for the gas to escape, it would have exploded the tank. The human effects then might have been greater or less, depending upon how much the pressure and heat would have diminished in bursting the container. The gas would have spread fast and low, killing many plant employees. Bursting water and gas pipes would have stimulated a continued vapor reaction. On the other hand, the explosion inside the tank might have exploded the other tanks and its cloud might have been distributed in all directions at first, and then spread over the whole south hemisphere that is, over the whole city.

A police inspector had alerted the police quickly, but the efforts of the police to alert the population ahead of the cloud of gas were unsuccessful. Phones aroused some and loudspeakers others, but both were most effective in the better residential districts where the gas was taking its smallest toll. Moreover there was some question as to the advice to give generally. The specific advice had to put the decision upon the individual and the individual had a hard choice to make: If you think you can evade the cloud then flee; if you think you cannot, close the windows and door and at the first sign of gas douse yourself in water and bury your face in wet cloths. (I would point out that in hot countries, houses are designed to admit outside air always, and anyhow most of them leak air.)

The police then joined in transporting victims to hospitals and burial grounds, for it was decided that the bodies had to be disposed of rapidly to avoid danger of plague and because there were too many victims to bury or burn ceremoniously. There seemed little likelihood, also, that identification could be accomplished in a large minority of cases. Looting was not much of a problem since practically everyone was in distress.

The Army Sub-area Commander was contacted by an ex-Brigadier General who was President of the Straw Products Company, with 176 employees working the night shift; they were assailed by the gas, and telephoned to him. The Army sent relief trucks promptly but many of the workers were stricken and some died. About 2:00 A.M. the Army Area Commander was reached and sent out a fleet of trucks that reached the devastated area within an hour. There began for the military a grim ordeal over the next few days of combing the houses for the dead and surviving. Other soldiers served as orderlies in the hospitals. The main activity now was in the hospitals, and can be described along with the several medical issues in the next chapter.

At Union Carbide's elegant headquarters in Danbury, Connecticut, away from the congestion and confusion of New York City and ten and a half hours behind events in Bhopal, the news brought shock and disbelief. No one was denying responsibility, or that the plant was part of themselves. The principal officer for environmental affairs assured interrogators that MIC was hardly lethal, rather a form of riot gas causing tears and coughing. Announcement was made that experts would be quickly despatched to the scene from America. The Chairman consulted his conscience and in the face of opposition from his staff decided to leave for India and Bhopal. Diplomatic channels obtained for him a promise of entrance and exit. Meanwhile the company's public relations staff was stepping up its operations. Scores, if not hundreds, of "communications specialists" were hired to contain public hostility and alarm. They answered thousands of inquiries from the public, politicians, and the press. News briefings were frequent. The defensive instincts of the Company were manifesting themselves; orders were passed around the world to refer all inquiries to appropriate authorities and to allow no visitors on UC premises anywhere without special clearance. The Company began to dig itself in for a state of siege.

When the UC Chairman flew to Bhopal on Friday, December 7, he was met at the airport and arrested by police. He was charged with criminal negligence, released after several hours of detention in the comfort of the company guesthouse, had bail posted for his release, and flew in an Indian government aircraft back to New Delhi. There he visited the Indian Foreign Secretary, under escort by the U.S.A. charge d'affaires in an Embassy automobile, and spent the weekend before returning to the United States. He had decided against a news conference in India because of the expected hostility of the Indian press, a warranted fear, inasmuch as he had only a modest ameliorative program to offer and could hardly speak up proudly or even informedly of the events at the Bhopal plant.

The Chief Minister -- Americans would say "Governor' -- of Madhya Pradesh refused to meet with the Chairman and in fact had ordered his arrest along with the top officers of the Indian Union Carbide company. It was understood, however, that his safe return to America had been promised by the Union Government. An offer of a relief grant from the Indian company amounting to ten million rupees was announced coincidentally and was refused. The Chairman's visit was not then a success and the Chief Minister scored in political points; however, not to have come would have appeared to be a heartless omission.

The State government was relatively inert, following upon the arrest of the Union Carbide officials and the seizing of the plant. Before then, forty hours after the disaster, a meeting of secretaries and heads of departments was called to coordinate emergency activity. A State Relief and Rehabilitation Committee was set up under the Chief Minister, which fissioned into two Committees. One, on finances, was to survey the damages but was overshadowed by the more ready Union Government relief payments. The other decided to distribute free milk to children and nursing mothers.

Besides preferring criminal charges against the managers, the State set up a Commission of Inquiry headed by a Justice of the High Court to investigate "into the events and circumstances of the accident, the adequacy of steps taken by the factory authorities, the adequacy of safety measures and their implementation in regard to measures for prevention of similar accidents in industries of this nature." The State has also served notice to Union Carbide of intended cancellation of its license to operate under the Insecticides Act of 1968, and the Chief Minister said that the plant would never open again. Further activity was expected upon the occasion of suits by the State against the Union Carbide interests in India and America, and upon the convening of the State Assembly, when hospital appropriations, industrial safety and rehabilitation measures would be taken up. At first the State denounced the American lawyers and then it began to sign up its own clients, until now 6000, and in March, with the Union Government, entered the U.S. Courts alongside the lawyers.

The incapacity of the State in the emergency is understandable. It possessed little in the way of equipment and few personnel trained for crisis management. Its own employees dispersed for their own safety. There exists no large welfare apparatus devoted to the problems of the poor. Confronted by a parade of demonstrators at the beginning of January, organized into a "People's Movement" to seek relief, the Chief Minister could only seek to belittle their pressure tactics and at the same time to placate them with assurances of concern and help to come.

It needs be said that underlying the unrest in regard to the State and City governments was the feeling, rather more widespread in India than in the United States, that both the political and administrative branches were ordinarily corrupt. After the excuse of over-population, corruption is the most common reason given for the painfully slow progress of India toward providing its people with a decent existence. The same State Government, the press was quick to point out, had been warned in advance of potentially grave accidents likely to occur on the Union Carbide premises. The factory location had been defended by State officials both as safe and as too expensive to move. This was in December 1982. No one knows, but many suspect, that damaging facts will come to light in the courts when the record of inspections and the inspectors themselves are brought forward. It is possible that required inspections were not performed, or were performed in a perfunctory manner, and that known defects in safety systems will have been passed over, and further that inducements of various kinds may have been traded.

The Government of India was similarly handicapped in responding. The Prime Minister, like the State's Chief Minister, behaved as proper politicians should; their sincerity and sorrow were genuinely felt and appreciated, just as was mother Teresa's. The Central Government within days was distributing adequate quantities of wheat, rice, oil, sugar, and milk to residents of all the affected areas of the City. It also announced that it would pay RS. 10,000 (about $800.00 U.S.) to every heir of a deceased victim, Rs.2000 to the seriously hurt, and 100 to 1000 to those less affected. These were called ex gratia payments, meaning free and unqualified, but serious problems arose in the course of paying out Rs.36.67 lakh (about $300,000) to 5724 victims, and led after four days to a suspension of payments. Prompt resumption of payments was promised but only in mid-January and after some talk of popular demonstrations were some payments made to the heirs of the dead. The problem of verification and administration was too great for the Government. The People's Movement was outraged. Why, they asked, should one hesitate to pay benefits on demand to the applicants from heavily affected areas? How, asked the bureaucrats, can we pay out money without proof? Some might cheat! They were using a figure of 1404 deaths, so announced by the State Government. No doubt the Union Government is feeling acutely the problems of determining personal damages and norms of compensation.

The Government of India, too, is planning to sue Union Carbide both in India and in the United States on behalf of the victims, and the State and Union Governments have in fact proceeded to join their suits in America. It is moot whether this action would deprive the victims of their individual rights to sue under Indian

Constitutional Law, and under the American Constitution. To an individualist and libertarian it is ironical and delinquent parents suddenly appearing in loco parentis to represent their neglected and victimized children.

And, of course, the Indian Government has set up a Committee to study comparatively the methods of regulating hazardous industries around the world. Initial attempts of the Opposition to debate the issues of Bhopal in the Parliament were frustrated. Nor has the Opposition, severely reduced and divided following the recent elections, managed to establish a non-Governmental position respecting the Bhopal disaster.

If all of this seems a poor response by the collectivity of concerned institutions to a vast tragedy, we may console ourselves with several thoughts. First, it is doubtful that the immediate response to such a disaster in other countries would be markedly superior. Second, the crowd of destroyed and sickened people pulls itself together and heals itself in a rudimentary manner and instinctively, individually and by mutual aid, apart from the institutional machinery of the government, which claims in this modern age a total and glorious mission that as often as not it fails to perform. Moreover we have not yet delt with the medical response, which was heroic. Fourth, the story is not finished; the possible outcomes of the devastation and decimation have not been related.

Finally, consider the victims, thousands in number arising out of the poorest of the Earth. Their own glorious annals deserve volumes. I cannot recount them here. Consider whoever was controlling trains at Bhopal. With people all around him stricken and crying out in agony and while others fled, himself half-blinded and hardly able to talk, he calls to the railway stations to the North and the South, telling them to stop the trains, keep them from entering the poisonous cloud at Bhopal. He died.

CHAPTER III

The Medical Emergency
and the Prognosis

All five hospitals of the City became crowded quickly. The first victim staggered into Hamidia Hospital at 1:15 A.M. and within an hour thousands appeared. Hamidia was the largest of the five hospitals and is joined to a medical school and research center. The beds filled everywhere; then there were two to a bed. The prior bed tenants were bundled off to their homes unless gravely ill. Then the floors were lined with mats and no one could move about without difficulty. So tents were set up on the grounds.

Doctors, interns, students, nurses, orderlies, friends and relatives of victims, then volunteers, all threw themselves into the seemingly endless work. Soldiers and police joined them soon. The doctors who received the first patients had to improvise the therapy. The event and hence the symptoms were totally surprising. What is to be done with eyes that were struck as if rubbed with hot chillies, with choking so severe that patients strangled before your eyes?

The Union Carbide India resident medical officer evacuated his family and turned his attention to the medical scene, which he found in full frenzy. The superintendent of Hamidia Hospital is quoted as saying ; "Probably sometime around 3:00 A.M. Dr.- -, when contacted after repeated attempts by one of our doctors, only told us to give wet cotton pads to soothe the victims' eyes-- something which is commonsensical, and which many victims had resorted to, before coming to us."

For the eyes simple washing in water was advised, then atropine drops to enlarge the pupils, this later only after someone reported poison gas in waves and the idea that the cornea might crystallize occurred. But when the patient could not see because of the atropine it was difficult to explain in the confusion that this was supposed to happen. Drugs were used to suppress the bronchial and lung inflammation and to dehydrate the body so that its own fluids would cease to flood the lungs that drew upon them avidly. But the patients suffered an agony of dehydration for they were already complaining of the fever and heat that were produced by the low flow of fluids into the lungs. Other drugs were used to dilate the bronchial tubes. Antibiotics were dispensed freely in anticipation of the infections that might follow the destruction of throat and lung tissue.

The damaged respiratory system could not take in ample oxygen. Respirators were applied. Oxygen was soon in short supply. So were respirators. Masks and tanks were brought in and flown in, so that the supply of these became adequate, but some say that the supply became adequate because oxygenation proved ineffectual. Lung suction machines were flown in from Britain and France. All of the while untreated patients and inadequately treated patients were dying or worsening.

Medications were abundantly available, if one could agree that the medication being administered was appropriate to the disease. Hospital conditions were abominable, a word that most foreigners from the West would have used even before the disaster, and it is a situation that prevails all over India. However, Bhopal boasts a medical college and therefore a supply of competent personnel at all levels. Some 853 doctors plunged into the massive task. Some of them came to the hospitals in the first place because they were personally affected by the gas and sought therapy, and then they stayed on for several days of exhausting work.

On the seventh day, a City Councillor assaulted a senior doctor, after failing to persuade him to retain a patient for the five days necessary to qualify for a benefit of Rs. 2000. The doctors went on strike; a placard read "Congress Party muscleman-ship is deadlier than MIC;" they also got some needed rest. Soon, appeased, they returned to work.

A thousand volunteers were organized by civic groups and welfare organizations, it is estimated, or came as individuals to perform the work of orderlies, to set up beds and fabricate stretchers from cloth and sticks and rope and to cut and sew bed cloths. Everyone had to conclude: there is no specific, no antidote for MIC poisoning. That the chief medical officer of Union Carbide India and the plant manager were unprepared for one case, let alone a quarter of a million cases, of MIC poisoning seems to have become evident in the beginning of the crisis. "An irritant," "a potent tear gas," these were the type of phrases applied, not only in India but also in the United States. When the dead began to pile up, they could not believe their own knowledge of what the tank contained. They wondered aloud whether some strange contaminant or reactant had somehow found its way into the tank and had proved to be highly toxic.

In the fatal hours inquiry was made to the parent company in America and the following telex message was returned the same day, December 3, to Union Carbide Ltd offices in Delhi, Bombay, and Bhopal. (Remember that the USA was running 10.5 hours behind Indian time.)

From Dr.Avashia, Medical Director-Institute Plant :

Urgent pass this info to doctors in Bhopal.

1. Treat patients with respiratory problems from methyl isocyanate by intravenous injection of hydrocortisone or prednisone 1 gm. immediately and after 24 hour. After also give oxygen and supportive treatment.

2. If cyanide poisoning is present administer sodium nitrite and sodium thiosulphate : if the patient does not respond to the amyl nitrite administration, or if severe exposure is suspected, administer intravenously 0.3 gm. sodium nitrite (10 ml of a 3 percent solution) at the rate of 2.5 to 5 ml/minute, followed by injection of 12.5 gm. sodium thiosulfate (50 ml of a 25 percent solution) at the same rate and via the same needle and vein.

3. Observe patient: The blood levels of methemoglobin should be monitored and not allowed to exceed 40 per cent. The patient should be kept under observation for 24 to 48 hours. If signs of intoxication persist or reappear, the injection of nitrite and thiosulfate should be repeated in one-half the above doses. Even if the patient appears well, this second injection may be given two hours after the first for prophylactic purposes."

Later on Dr. Avashia said that he had gotten his information from his wife, who had heard a radio broadcast about the disaster in which the theory that cyanide poisoning might be involved was reported. Therefore he inserted the advice on cyanide. This message thereupon added itself to the local rumor at Bhopal of mass cyanide poisoning, which was generating out of the severity of the symptoms and the suspicion that some new unexpected horrible gas had been released.

We note also, concerning the telex, that the third paragraph refers to the treatment for cyanide but might be construed as treatment also for methyl-isocyanate poisoning, and this too contributed to heated arguments among the attending doctors over experimenting with the cyanide treatment on the MIC patients.

Finally, concerning the telex, its language is obviously intended for the care of only several patients, not a mass poisoning. Practically none of the scores of thousands of victims received the attention here prescribed.

In mid-March, the Indian Council of Medical Research announced preliminary results of a study of persons exposed to MIC, who typically had high levels of thiocyanate in their urine. In a double-blind test, half of a sample were injected with thiosulphate and half with a placebo. The experimental group showed symptomatic improvement and an increased excretion of thiocyanate. This excretion was known to occur in cyanide poisoning and hence now to be indicative of MIC poisoning, for it implies a breakdown of MIC into cyanide. Thus, what was hotly argued by the doctors in the middle of the disaster and rejected -- that thiosulphate was an antidote to MIC -- now appears likely.

The symptoms of MIC poisoning have been familiar around the world and anyone who goes into the shanties of Bhopal can get first-hand accounts of them. MIC is a common element in pesticides and pesticidal poisoning is endemic everywhere. Yet there must be thousands of public health and medical officers around the world who have observed victims of poisoning without bringing effective pressure upon the sources of the poisoning to investigate and publicize what they know about the materials they are producing and shipping. In this case some fault must lie with the parent company and the U.S. agencies that are charged with research and dissemination of information in this area. Since the United States Congress votes the largest appropriations in the world for public health research, it would appear in order for a committee of Congress to investigate the failure on the part of both Union Carbide and the several Federal agencies concerned to live up to their responsibilities.

If the first symptoms of MIC poisoning are corroding lungs, inflamed bronchial tubes and throat, glazing eyes, and gastrointestinal upheaval, they are not the complete and final list. Comas, indicating a trauma of the central nervous system, were frequent, also dizziness, and often severe muscular weakness, again denoting neurological disorder. Postmortems were conducted in many cases. The lungs were found to weigh twice or three times the normal because of the weight of fluids in them. The blood of a patient who died on the first day was pink; if he died after a week of 168 hours of agony, his blood shows up a dark red and his organs are congested. MIC or its compounds, amines and cynamide, are found in the blood. The hemoglobin of the blood is deactivated.

Major questions of cure arose. That people continued to die settled the point whether the disability might be progressive. Several were dying each week, months later. For how long would the possibility of death be imminent? And, if not death, how long would the disability persist? Union Carbide USA arranged with well-known specialists, one on pulmonary conditions and the other on ophtalmology, to fly to Bhopal a week after the gas escape. Both gave optimistic statements after three days of examining patients. MIC, they said, does not course through the body and affect organs other than the lungs and eyes. Once past the initial crisis, the patient's bodily fluids will dissolve the MIC. However, pneumonia and bronchitis may more readily develop out of lung damage. Furthermore, in cases of prolonged oxygen deprivation, the heart, brain, nervous system, and, with pregnant women, foetuses, may suffer damage. Permanent eye damage was discounted; any damage would repair itself except where the cornea had become discolored and this would, the doctors promised, be repaired by corneal transplants.

The treatment afforded most victims was a matter of a few minutes. They then dragged themselves back to their hovels, thereabouts to remain to this day, three months later, mourning their lost ones, complaining of their physical weakness, trying to keep track of their periodically renewable medical prescriptions, and venturing out or sending someone to get their food rations. Mobile clinics go about dispensing liberally analgesics, antibiotics, eye washes, and antacids, arranging also, in cases of pathological changes, for hospital examination.

The deep mental depression that has settled down upon many thousands of victims goes unattended. This is not regarded as an "illness." No one has yet alluded to other mass traumas and counseled: "See here, you have a mass trauma as bad as those of the Nazi camps and Hiroshima. This requires as many doctors and clinicians as the physical trauma." And, of course, the physical and the mental are interacting. Despite all assurances and display of caution by the authorities and experts, when it came time to process the remaining tons of MIC in order to dispose of it safely, a terror seized the city and a massive evacuation took place. The officials and scientists in charge of the processing went to ridiculous extremes in providing safety measures, giving the pompous title of numbered safety systems to such impressive expedients as hanging wet cloths over the gas vent, draping wet cloths over the wire fences, arranging for extra fire-fighting and water-hosing equipment to stand by, and sending a water-laden helicopter into the air to flush any vapor cloud that might be aloft.

The "People's Movement," in close contact with the victims, realized their psychic state and brought forth as well the variety of physical symptoms that were both cause and effect of the mental ones. Besides breathlessness, burning of the eyes (it was painful to go out in the sun), impaired vision and coughing exhaustingly, there were adduced the common complaints of bodily weakness, bodily pains, palpitations, giddiness, frothing at the mouth, convulsions and comas. Long after the first week, people returning home felt dizzy and even fainted upon sweeping floors or opening doors and cupboards. That MIC clings to cloth and rugs an American reporter from a chemical journal discovered; all dry goods had to be well washed. To many the well water had a metallic taste. The City water and the lake waters were pronounced safe to use, but few believed the announcements, and the fish and meat and vegetables coming on the market were likewise shunned or eaten fearfully despite pronouncements of their edibility. It was also discovered that people who had been sitting by their smoky little fires or who had been drinking alcohol before the gas cloud crept in seemed to have been less affected by it.

The "People's Movement" registered some botanical and zoological observations as well. Leaves of all types of plants and trees were scorched to a blue-black except the tamarind (imli); pomegranate leaves turned yellow, pipal leaves black. Chickens were less affected than fish, goats, cows and dogs. The upper layers of stored wheat flour (atta) took on a greenish hue. The hundreds of animal carcasses, most bloated and near bursting with their gases, were finally picked up by hand or by crane and dumped in a freshly dug, one acre-square ten-foot pit and spread with salt, bleach, lime and caustic soda, then covered over with earth.

A month following the visit of the American doctors who gave an optimistic report, a McGill University pharmacologist came to Bhopal at the invitation of the People's Movement and examined about fifty patients. He condemned the governments for not providing public reports, even preliminary ones, on the pathology of and therapy for MIC poisoning, and he termed the pronouncements of the American doctors to be "outrageously unscientific." He commented upon the recurrent breathlessness and conjunctivitis, extreme weakness sensed, and loss of appetite and taste, the abdominal swellings, and the discoordination of physical movement. Further he stressed the severe disruption of menstruation in many women, and the occurrence of spontaneous abortions. It was elementary, he claimed, that when a poisonous gas arrives at the lungs, it proceeds to other parts of the body, and that the liver and gastrointestinal tract were likely to be affected. Despite much evidence of internal bleeding, no endoscopies were being performed because of a lack of instruments in working order. Mutagenic effects might take eight to ten years to detect. Finally, there was a total lack of any rehabilitation programs for the victims.

An equally critical account was published at the same time in the British journal, New Scientist. The lack of a long-term blood-testing program at Bhopal was deplored. A chance to learn of enduring and progressive effects of exposure was being lost; after January the blood levels of MIC in many persons might not be detectable. If blood tests were to demonstrate that MIC bound itself to hemoglobin and nucleic acids, mutations and possibly cancers might occur in the absence of countervailing mechanisms. Any highly reactive agent like MIC can react with proteins and DNA in the cells to cause cancer. Widespread metabolic disorders are possible, given the propensity of MIC to combine with amino acids and proteins.

Shortly before the Union Government appointed a committee of his group to investigate the disaster, the Director-General of the Indian Council for Medical Research declared, or so it is reported, that "there is no reason to believe that there will be any long-term effects of MIC poisoning on various organs and systems of the human body." Later his deputy vigorously denied that such a statement had been made. Nevertheless, this opinion, favorable to whoever wanted to underplay the events, seems to have preponderated for some weeks and still is prominent in interested circles of India and USA.

Yet the Indian Council's preliminary report of December 15 did state that "severe tensions would be expected to lead to significant scarring (of the lungs) in the course of healing amongst long term-survivors. Pulmonary fibrosis of a bilateral nature is expected to ensue with its attendant effect upon lung function. Careful contemporary and follow-up studies are needed not only to evaluate the nature and extent of the residual damage but also to search for any clues in the management of these cases in the short and long term to minimize residual damage. With the appearance of sub-acute and chronic syndromes as time passes, it is expected that target organ damage, if any, other than pulmonary damage, might be expected to become increasingly evident." The Committee set up a research unit to coordinate the work under the Dean of the Gandhi Medical College at Bhopal.

In a press conference at Delhi in February, the Director General announced that postmortem reports had revealed the presence of cyanide, that the respiratory systems of the cases examined had undergone "devastating changes", that a "significant and striking feature" was the cherry red color of the blood in all the organs of the body, and that there had been brain oedema, neurological disorders, degeneration of the liver and kidneys; further, lungs had doubled or trebled their weight. In confirmation, later deaths and illnesses occurred in conjunction with fat excesses in the liver, degeneration of kidney tissue, spleen injury and gastric and intestinal ulcers.

On March 21, an infant was reported born of a mother who was resident of an affected neighborhood of Bhopal (Jehangirabad) during the gas cloud release and who had suffered vomiting and burning eyes. She left Bhopal four days after the disaster and the baby was delivered at the Bilaspur Hospital in Raipur. The baby weighed nine pounds, came after what seemed to be a normal pregnancy, and was delivered by operative assistance. However, the infant's eye cavities were void; its skin appeared scorched, its fingers and toes were undeveloped; and its sex was indeterminate. It expired within forty hours. A pathological investigation was ordered to seek signs of MIC effects.

Under such circumstances, it would be premature to claim that permanent damage would not be suffered or that new symptoms would not occur or that injuries already evident would not worsen as the poison's effects gradually took command in various organs or that there was no possibility of birth and genetic problems. The resolution of the total complex of issues cannot be well considered, therefore, without some provision being made for the appearance of new symptoms or the worsening of old ones.

The heavy toll of the disaster and the blackout of company and government news on many aspects of the case has led to various conjectures about the substances employed, the large quantity of MIC on hand, etc.; at some point these must be investigated, even though they will be most likely dismissed as false. Union Carbide built a few years ago a research structure on the factory site at Bhopal. It is remotely possible that the research facility was being used or intended for use to test the chemical warfare potential of MIC or to develop other chemicals that would be hazardous in themselves or when compounded. Indian journalists have raised such issues, and have found a large audience receptive to the theories, despite a denial by the Indian Ministry of Science and Technology that it had authorized Union Carbide of India to undertake any research related to chemical warfare; the Bhopal facility was merely one of 900 laboratories in India that had been granted incentives. But, think the suspicious ones, the CIA would have been in association with the Company and certainly not reporting to the Ministry.

There is widespread suspicion of American motives and conduct in international affairs. Hence it may be well to open up the full record of the motives, economics, decisions, and correspondence that led to the founding of the research center. In the aftermath of Bhopal a not uncommon view is that voiced by the tabloid weekly, Blitz; it wonders whether Union Carbide will be permitted to escape full responsibility owing to the fact that the victims are Indians rather than Americans, and it criticized American law that gives compensation to U.S. servicemen who have suffered from their exposure while employing the defoliant "Agent Orange" in Vietnam, but offers nothing to the Vietnamese civilians who suffered from the same poison in much greater numbers.

The most effective mechanism for examining issues such as this would be the U.S. committee system in Congress; it can call hearings, order documents brought before the committee in charge, subpoena witnesses and eventually dispose of the issues.

In the tens of thousands of words of reportage and in all the eyewitness accounts of the events at Bhopal, one reads or hears of nothing that would constitute a systematic attempt at discovering the numbers of victims. Nor does it seem that anybody was in a position to make such an enumeration. At the hospitals, record-keeping is ordinarily done in a less than perfect manner; The Times of India recently published an investigative report documenting the lamentable state of medical records. During the emergency all semblance of control over admissions and releases was lost. Consider only that patients by the thousands were camped on the grounds outside. All control over burials at the Muslim cemeteries and of the mass cremations by the Hindus was lost. (A man, thought dead, climbed down from his own funeral pyre.) There was no accounting for the numbers of people leaving the city and returning.

The moment came when serious estimators were trying to arrive at the figures of the dead by guessing the amount of wood that was used in the fires that the Hindus used as funeral pyres. (The State Government later announced with macabre pride that its Forest Department had provided 20,000 quintals of wood, two million pounds, for the crematory holocaust.) Nobody knows how many bodies were cast into the waters. One small boy pulled himself from the waters into which he had been tossed as dead. Many bodies were carried to the villages and as far as Indore for burial or cremation, and many persons died too in these removed places.

The Delhi Science Forum sent in a team to visit the City shortly after the accident and estimated the dead at five thousand, the seriously injured at fifty thousand, of which 20,000 were in critical condition. The number of persons affected by MIC was guessed to be 250,000.

For death-gift purposes, as I have said, the Government of India chose to adopt the figure of deaths supplied by the State Government, about 1400. The Tata Institute of Social Sciences, with collaborating groups, was asked to survey gas-affected families. In mid-February a preliminary report, covering 25,294 families residing in 36 neighborhoods near the plant, confirmed 1,021 deaths and 1,064 cases of blindness. One hundred and fifty children were orphaned and 168 women widowed.

The survey was conducted by questionnaire, mostly in a fixed framework of response, and administered by paraprofessional investigators. Although the most extensive attempt to solve the problem, it raises some questions. The number of families appears to be about half of the people known to have been affected and treated in the hospitals apart from the dead. The number of blind in proportion to the dead seems excessive unless the dead are unreported. Orphaned probably means deprived of both parents. Given the 168 widows, of whom are the 1021 dead composed? Were only 168 of them married men? The many families and individuals who were erased from existence, or who permanently abandoned the area are not accounted for. In India, widows especially would be forced back to their villages. Our working estimate of 3000 deaths seems to hold up unless queries can be handled convincingly in a true, full report by the Institute. Furthermore, the field investigators were forbidden to make tallies, and the methodology has been concealed, an abuse of social science itself calling for investigation. The

State government is sitting on the data, and the Tata Institute appears helpless to dislodge it.

A principal medical doctor and chief of section at Hamidia Hospital, who was in the thick of the holocaust, agrees with our estimate of 3000. Circumstantially, the burial procedures would support it, too. And if one examines the press accounts, they are observed to begin the first day with estimates of around 600, to move to 1000 the next day, and then to climb to the vicinity of our estimate before leveling off. They are probably more reliable that the Tata Institute figures, which followed in time and numbers the State government estimates adopted by the Federal Government, and were derived some weeks afterwards by methods that are not entirely clear.

When Bhopal voted, along with several other parts of India, in a special delayed election, six weeks after the disaster, participation was at least 25% below normal, 40% as contrasted with an average of 65% in all other areas voting. Where had this 100,000 or so people of voting age gone? Perhaps they were still refugees. If so, add another 100,000 for the young, making 200,000 semi-permanent refugees. To this day some persons insist that many more died; before one discusses such claims, it is well to recall how long it took to discover how many had died in the Nazi holocaust, and how "right-thinking" and official opinion was usually on the reductionist side.

CHAPTER IV

Secrecy, Reductionism and the Press

The first impulse of Union Carbide (USA) was to tell the truth, namely, "We know nothing." But the press and public will not accept "nothing" for an answer and will start to compose their own version of events. They ask about the designs of the plants in Bhopal and around the world, about specific personnel, profits, past decisions, etc., until it is quite obvious that "nothing" is actually "much", or at least a good beginning. At this point (and we are speaking of a day or so, or even hours) Union Carbide begins to think in terms of secrecy. "Anything we say, and that means information, too, will be used against us." So secrecy is imposed on all officers and employees, to the extent possible.

Enter the attorneys and public relations staff and matters are made worse. Secrecy is a direct function of the sum of public relations and legal talents engaged. The more of one, the more of the others. The pressures to know cannot be resisted and a press conference near, not at, company headquarters in Danbury, Connecticut, and a tour of the West Virginia plant, the "twin" of the Bhopal installation, are arranged for the press. The media and public relations element in both arrangements prevails. The *New York Times* complains politely of the excessive caution, the restrictions imposed on camera crews, the failure to answer practically any questions of import. Thus the West Virginia plant is declared safe but its operations were not describable, it seems. A computer system collects information, but how and for what is not stated. Surprise was manifested, too, when the ability of the system to handle a Bhopal-sized vapor release was questioned. Nor would the guides tell reporters whether automated safety features, purportedly present in West Virginia, were incorporated in the Bhopal operations. A hasty Congressional hearing brought out expressions of sympathy and concern for safety by the UC Chairman and others, but little hard information.

The security syndrome spread, and one could witness the next phase of secrecy shaping up, whereby a direct relationship occurs between the extent of secretiveness and the growth of self-deception and fear on the corporate side and the negative attitudes towards the "bad guys" on the press and public side. Also one sees an evolving relationship between secretiveness and the slowness of response all-around (legal, corporate, governmental, voluntary) to the needs of the victims. And further, as secretiveness is pursued, active help from outside quarters and constructive ideas of responding to and resolving the crisis are suppressed.

The U.S. corporate leadership is not alone in secretiveness and suffering the effects thereof. The State government arrests the Indian corporate leaders and prevents them from agreeing upon a story with their U.S. counterparts. (Clever, but perhaps they might disagree!) The Indian executives become incommunicado and uncommunicative, resentful at their arrest. They are finally released, and of course communications with the U.S. side resume. What can they be saying? Not much, I believe. It is not easy to agree on such a big story, and the motives for disagreement are as obvious as for agreement. They are not all in the same boat.

The State government, City of Bhopal, and the Indian government also impose secrecy on all whom they command. They shut down the plant as well, opening it only to exhibit the processing of the remaining MIC. The governments even engage the doctors and hospitals in the game. These are instructed to repel reporters seeking news. The Chief Secretary says that the government did not want too many medical opinions floating around, for they created confusion and apprehension among the population. However, just the opposite occurred: the more the secrecy, the more wild rumors circulated and were printed.

The results of all this secrecy are not promising. Victims find their cases at law deteriorating because they can find no one to make determinations of their original condition and describe its progress or decline. They cannot produce a blood test to show that their blood did or did not contain MIC. Their cures also lapse, because, as secrecy increases, treatment and issues of therapy become undiscussable. S. Khandekar, writing in *India Today*, reported that the Secretary of the State Legal Aid Board was handling with alacrity the rush of people to fill out forms for the State to pursue their legal cases, but had little or no regard for the need for medical documentation in these cases.

The public is not given a steady flow of information to maintain its equilibrium. It is unable to learn the lessons of Bhopal and to demand that they be applied domestically and throughout the world. Voluntary public efforts are discouraged, because the governments and corporations imply that these are misguided or unnecessary "if only they knew the facts."

Science suffers, too. A petition by a Bhopal citizen, presented by his lawyer before the High Court of Jabalpur, asked that the State Government be directed to preserve, in its operation to detoxify MIC, a sample of the contents of the tanks and to not tamper with the equipment employed at the plant. The Judge did so direct that 15 kilograms be kept in 3 containers and enter the custody of the court at Bhopal. Union Carbide (India) and the State opposed this, asserting the procedure was not feasible and was dangerous. The Judge reduced the amount to 1.5 kilograms. Still the Company and State argued that it was not safe. An appeals court of two justices upheld this amount and directed that the sample he analyzed by the chief government scientist on the project and a scientist of the petitioner's choice.

The Indian Council of Medical Research advised individual investigators not to publicize their findings prematurely and individually; confusion and panic might thus be avoided. It favored collating reliable and authentic findings for communication to the public and medical profession. This statement would indicate, to the critical mind, that victims would have to wait a long time and offbeat findings would be exorcised if the 'scientific process" as defined here were accepted.

The net gain, looking at secrecy from the standpoint of the corporate executives and politicians, is nil, and quite possibly a net loss. Besides being ineffectual, misleading, irritating, and presumptuous, it prevents the kind of creative basic rethinking of the problems that should be inspired by the immensity of the tragedy and its implications. If the heads of Union Carbide in both countries and of the governments let the processes of free discussion operate, better ideas for resolving all the issues would occur to them, and they would be helping the total cause of freedom of information and free political systems in the world.

The press, both American and Indian, did a creditable job of reporting and analyzing the Bhopal disaster and its aftermath. It continues to do so, despite the curtain of secrecy dropped quickly over the case by the authorities and maintained in place to this day. Coverage has dropped because of lack of resources, because of the secrecy, and because to say anything more one would have to venture into vast areas not defined as "news." To mention only several, the *New York Times*, the *Times of India*, and *India Today* have presented extensive material of high quality. That the *New York Times* will ultimately have spent half a million dollars to treat the Bhopal disaster is in line with the serious nature and ample resources of that journal. The reports of volunteer groups, quite another genre, such as the People's Movement, the Delhi Science Forum, and the Eklavya group also contributed valuable services.

At an extreme from the *New York Times* in every material regard would be the *Hitavada*, a diligent newspaper published in the English language in some 12,000 copies a day at Bhopal. Here one can find the archetype of so many American stories and films about a bygone day, authentic folk heroes, the fighting editor with the couple of reporters, operating in a corner of a cavernous cement room where hand fonts and antique linotypes feed composition to loosely clacking presses under pale weak lights.

What has the *Hitavada* done? It has covered the great story from hour to hour with profound compassion and solicitude for the victims. It has been a troublemaker for the authorities and experts, pointing out many contradictions without regard to party, asking probing questions (sometimes far-fetched), venturing pessimistic opinions, and pressing for medical help and compensation.

Another dimension of the press in Indian society, as in America, is its symbiotic relationship with the voluntary sector. One provides news; the other publicizes the activity. Thus can the poor swing their weight about with some effect. Despite their seemingly hopeless problems of numbers and scarce resources, the Indian people energetically "petition for a redress of grievances" and act "peacefully to assemble," if I may employ U.S. terms. Environmentalists quickly responded from several centers, such as "The people's Initiative" of Bhopal, Zahrili Gas Kand Sangharsh Morcha. From several cities like Delhi and Ahmedabad, environmentalists responded with all too scarce resources and the U.S. groups were totally caught off guard; the charities (Red Cross, Catholic charities, *et al*) responded, but much response was of the "Just think if it had happened here" kind, and much of the Bhopal coverage in the U.S. news was actually coverage of the non-news that nothing was happening, and why a disaster was or was not going to happen at the Union Carbide plant in West Virginia.

"People's Movements" were generated promptly in India. When relief was delayed, street demonstrations were held. Film-makers were among their leaders -- "participant observation" would be the sociological term for it. Pickets marched before the government offices. Manifestations of a public opinion that would not have been heard, if left to the victims alone to voice, or to the governments or the corporations, were publicized in advance and afterwards in the newspapers. Authorities made light of the agitation, but with the press writing about it and about the governmental response, some help came fast.

Precursor to the thousand reporters who descended upon Bhopal after December 3 was a solitary journalist from Bhopal, Rajkumar Keswani, who in several articles beginning in 1982 attacked safety precautions at the Union Carbide installation as inadequate, and predicted a general disaster to the City. He began his one-man campaign in a weekly Hindi newspaper on September 26, 1982. On October 5, several days after another of his articles appeared, eighteen workers at the plant were injured by gas leakage. In November he wrote to the Chief Minister to admonish him. To this letter no response was received, he claims, but the office of the Chief Minister denies having any record of such a letter.

In a large Hindi newspaper, *Jansatta*, on June 16, 1984, Keswani published details of the highly critical American experts report of May. 1982, and warned that the Bhopal population could be wiped out. Further, he revealed that the leak of October 5 had sent thousands of residents from the nearby shanties fleeing, to return only after many hours of anxious waiting. (I would note that this incident demonstrates that the neighborhoods nearby did have an awareness that great danger was housed in the plant; still no action was taken by the company or authorities to explain anything to the people to advise them how to behave in a larger incident, or to ready the governmental and company personnel for emergency duties in connection with the community.

When one considers newspapers such as *Hitavada* and reporters like Keswani, and what would exist in their place if they were eliminated -- as in fact has happened in many nations of the world -- one comes to understand better the role of the press in forestalling the death of worthy causes; this occurs at the same time as, and despite the tendency of, the press to drop a story as soon as the story moves into the process of resolution and abstraction. Without the free Indian and world press, and despite the worthy pugnacity of the victims' lawyers from the USA the substance and meaning of Bhopal would already be markedly reduced. Carrying forward and dramatising the news, significance, and symbol of Bhopal, the press transformed the tragedy into the form needed if there was to be a full hearing, large help and illuminating history.

The same Rajkumar Keswani explained in the *Free Press Journal* (16 December 1984) the failure of his solitary battle against unsafe conditions at Union Carbide Bhopal:

Mr. (name withheld by the present author) former Inspector General of State police was employed by Carbide as security adviser after his retirement. This shielded them from the police. A Congress (I) Leader is their lawyer. The posh Union Carbide guesthouse was always at the disposal of the ruling party. A separate suite was reserved for Chief Minister----------. Mr. --------- used to stay there whenever in Bhopal. During the Congress (I) regional conference, all central Ministers were accommodated there.

Senior politicians and civil servants were obliged to the company for employing their sons and relatives on fat salaries. On its payroll are the nephews of former Education Minister---------and Irrigation Minister---------.

This symbiosis or business, bureaucrats, and politicians can hardly strengthen the technological heart of industrial enterprise.

Several forces operated to reduce the dimensions, meaning and treatment of the case. Hardly had the first deaths been reported when denials were generated concerning the scope of the accident, and of any possible negligence, whether of individuals, systems, governments, or companies. Most of this was motivated by self-interest. But even among the unaffected liberal public one met up with psychic denial, a collective amnesia at work, telling one "Let's not make too much of it." Attacks on American lawyers also tended to divert attention or reduce the issue by ridicule and displaced indignation. In several instances I observed well-intentioned friends switching their attention from the plight of the victims to righteous anger against the "bloodsucking lawyers."

Conservative reductions of the deaths, injuries and massive social dislocation were commonly encountered. Hardly disguised was the hope of some that by some mysterious means those who survived the disaster would wander home to their villages (which are presumed to exist as some Happy Home ready to receive their errant children). Then, too, some medical experts were joined by many others in reducing the gravity of the illness, and in exaggerating the hypochondriac behavior of some of the patients, who had, it must be remembered, deep psychological wounds as well as suffering and agony, such that, after being seen in thousands of cases, make even a sympathetic doctor wonder if some play-acting is going on.

In their exhaustion, the immense drama becomes surreal, just as it does for soldiers caught in the middle of a great battle.

Comparisons are made with disasters such as Hiroshima both to reduce and inflate the importance of Bhopal. Some people wish right away to put Bhopal out of mind and begin discussing marvelous new "safe" pesticides and natural ways of fighting pests. Others wish, "Naturally, turn the matter over to the government involved." Some say that the Corporations cannot pay the damages involved and so mentally they reduce the allegations against the companies and thus the compensation foreseen. Others accuse the victims of being illegally in the path of the poisonous gases, of being "illegal squatters," as if they had no business existing or should have been on holiday at the seashore when the cloud came over Bhopal.

The poorest of people have at least the sensibilities of the well-to-do for sorrow. Grief among the well-to-do and the secularized gentry of modern times is often an arrogant feeling of being insulted; expecting very much from the world, they feel chagrined when the world seems to turn against them. By contrast, the very poor, already inured to insult and injury, grieve more sincerely than many of the rich for the loss of their loved ones. They have little in life besides their loved ones -- a few clothes, several pots, a goat, and sticks of furniture in a room that they can only hopefully regard as their very own. They have no job guarantees, no welfare system to speak of save family and friends; they may be allowed to lie on a mat on the hospital floor if seriously ill, or they will squat there tending to a dying relative. The ill now must live on, partially blinded, with coughs and weakness of the limbs, musing upon the dead. Even the workers of Union Carbide, once they will be left go by Union Carbide, will become the instant poor, living on next to nothing and flooding into the crowded slums.

Hope can actually rise in the breasts of some people as they announce their happy discovery that the victims were largely the poorest of the poor and were almost surely illiterate, probably degenerate, and had too many children, and so on, just as they would have spoken of Jesus Christ, that "he was probably well crucified since he owned nothing but a cloak," which, we might as well add, even that was stripped from his body and parceled out.

Another view found in American circles, is that the Bhopal tragedy was a Indian affair, the Indians will botch it, they will settle cheaply and, them, paradoxically, at the same time they will say that the Indians are the only ones competent to handle it.

There is a constant tendency, among parties as diverse as the poorest of victims and learned American environmentalists, to reduce the Bhopal problem to a particular safety failure for which an assignment of responsibility and quick compensation are the proper resolution. To the contrary, I would say that the meaning of Bhopal needs to be preserved and enlarged. It is a jolting reminder of the gas ovens of Auschwitz, the radiation cloud of Hiroshima, the burned women shirt-makers of a New York City sweatshop whose death began a new chapter in the history of safety and better working conditions. Bhopal can be a watershed in industrial, even in world, history if the victims receive fair treatment and full justice, and if a new code of conduct comes to govern transnational business operations.

CHAPTER V

Who Sues Whom for What?

The venerable dean of torts, Melvin Belli of San Francisco, was foremost among a score of U.S. lawyers who descended upon Bhopal in the wake of the disaster. In February, legal suits, begun in several American states on behalf of the Bhopal victims, were all referred to the United States Southern District Court of Manhattan, New York City, this appearing most convenient to the litigants. Thousands of victims had signed delegations of authority for various lawyers. The number of counsel engaged in India is still very small compared to the Americans, of whom upwards of sixty have put in an appearance.

Belli's clients were not asked to sign contingency agreements but other attorneys asked up to 30% and expenses from whatever sum, if any, their clients might receive. (They do not really expect so great a reward for their services; the courts will ultimately be arbiters of a just compensation.) All of the suits named certain clients and added on all other similarly placed victims, that is, the whole class of persons affected. Such "class actions," now common in America but unfamiliar to Indians, if successful, will achieve awards for all those victims who are similarly situated with the named, represented victims. An enormous amount of repetitive litigation is avoided.

In India there exist doubts that a class action is permissible in the present case. The slowness of victims to file suits in India results from the awkwardness of processing such a mass of suits under the circumstances, from the resistance to contingency-fee arrangements, and from the high cost of litigating in Indian courts, where a plaintiff has to put money "up front" and may, if unsuccessful, be stuck with his attorney's fees plus court costs. If it were not for the fact that certain highly placed Indian leaders, the Chief Minister of Madhya Pradesh and the Chief Justice of the Supreme Court of India, have promised that all deposits and costs would be waived for a poor petitioner, almost none of the victims could afford to file. This act of charity and grace is just that -- and as yet a promise -- so that the victims are not streaming into the Indian courts for their justice. Perhaps the tragedy will bring about needed reforms of the Indian legal system in the direction of equality and personal rights.

It is in this light that one can reassess the outcry against the "legal vultures" and "ambulance chasers" that arose in the Indian and world press. A revulsion against the rush to seize upon sick and poor victims as clients is natural, even more in liberal and enlightened circles than among conservatives. However, let me suggest additional reasons why one should not hasten to condemn the American lawyers.

Who could be trusted to plead the victim's cause other than the American lawyer? The Indian courts and lawyers? -- But it was not until the American lawyers came stampeding that Indian officialdom was shamed and empowered magically to remove the Indian obstacles for these poor clients. Who else? Union Carbide corporations of the U.S.A. and India? The State of Madhya Pradesh? The City of Bhopal? The Indian Union government? The impoverished "People's Movement"? All except the last had their own axes to grind. They might well be brought to the bar on their own accounts.

Who were tightly and personally bound to the victims' interest in the highest and fastest compensation? The American lawyers alone.

Who also could plead the cause before the U.S. Courts? Only the American lawyers. Later, goaded by them, the Indian governments came to think that they might have a good case in law, too, before the U.S. courts and might be allowed to plead before them. Since the Indian governments are involved as parties to other aspects of the Bhopal issues, they may risk disqualification for not "appearing before the courts with clean hands."

Who would pay the victims' legal fees if they lost at court? Now the Indian governments will sue in their name, they say, and pay the heavy costs of litigation out of the lean public treasury. If their action is truly for the convenience of the victims, then they should consider these sums as advances to be repaid from the awards. Is it just to save money for the victims that the Indian governments intend to go to the U.S. courts? It must be so, because they can hardly do as well before the courts as Melvin Belli and others like him can, and there is nothing that the court will permit them to say that the lawyers of individual clients cannot say.

Furthermore, they have their own cases to fight against Union Carbide and other multinationals, in and out of the courts; victim litigation might conceivably be used to avoid addressing other issues which are very important to India and the world and arise out of the Bhopal disaster.

Who could better fight off attempts of a third party (a government, say) to determine and control the compensation, accepting less, charging for heavy administrative costs, and forever subjecting the victims and survivors to a bureaucracy, even if a benevolent one?

Who can better afford and more effectively fight to keep the case alive, and away from the compromises of India-U.S.A. relations? Who can better establish the case as a landmark of world law and justice on multinational industrial concerns? The American lawyers, I would assert.

And, finally, who can better force testimony from the American side and pry evidence loose from private, corporate, and governmental sources? Again the lawyers. If the governments are permitted to prosecute on behalf of the victims, one can foresee some prolonged legal disputes over the forcing of release of information from the Indian governmental side that will be embarrassing to reveal, as well as the calling of witnesses who will refuse to testify, on one ground or another, or perhaps who will find themselves pleading the Fifth Amendment (the right not to testify on grounds that one might incriminate himself) in an American Court!

Why were the American lawyers so eager to offer their services? One need not denounce their expressions of sympathy for the victims as hypocrisy. Suing a big company on behalf of a little fellow in so horrendous a case is a classic ambition of the American torts lawyer.

As far as concerned the liability of the company in the tort, the case appeared clear. Every student who has taken a first course in Tort law in an Indian or American law school has learned the opinion of the court in the case of *Rylands* vs *Fletcher*, heard in 1868 in the British House of Lords. There one reads that "the true rule of law is that person who, for his own purposes brings on his land and collects and keeps there anything likely to do mischief if it escapes, must keep it in at his own peril, and, if he does not do so, he is *prima facie* answerable for all the damage which is the material consequence of its escape." This is still good law.

The doctrine of strict liability, now established in American law, holds that a company may incur liability for escape of its hazardous substances causing injury. Prolonged investigation of causes becomes unnecessary; mainly the source of the substance needs be proved.

Courts have the power to determine the form that payments of damages may take, as a logical extension of the problem of dealing with international currency exchanges here, among other reasons, Moreover they can determine the period of time over which damages may be paid. In one English case, a young man acquired the obligation for the support of the full term of dependency of a family whose breadwinner he had negligently killed.

American courts, too, have great informal powers to manipulate, even when they cannot coerce, parties appearing before them into a settlement "out of court", at the same time indicating to them the kind of arrangement which the court would find acceptable.

In determining whether to accept jurisdiction over the cases brought before them by the Bhopal victims, the U.S. courts will ask themselves whether the Union Carbide Corporation centered in New York City with headquarters at Danbury, Connecticut, was responsible for the actions of the employees of Union Carbide of India, incorporated in India. The criteria to be applied will include the extent of its formal and effective control over Union Carbide India, in a chain of communications leading from the U.S.A. to the employees.

The formal control is manifest in the ownership by the parent company of 50.9% of the shares issued by Union Carbide India. With such a majority interest, the U.S. company can make, while abiding by Indian law, every kind of decision it pleases governing the affiliate, from electing members of its Board of Directors to determining dividends and salaries, not to mention requiring safe practices in the plants. This fact is so potent in law, that any evidence brought to bear by the Union Carbide defendants to demonstrate that they lacked real control is likely to be turned into evidence of negligence. If a man's dog leaps over his fence to bite a passer-by, the man's explanation that the dog taught itself to jump over the fence can only add to his culpability. For he should have de-trained the dog or built a higher fence.

It is conceivable that the U.S. Corporation might argue that its controlled affiliate was ordered by the Indian and or Madhya Pradesh and or Bhopal governments to conform to certain laws, rules, and orders that they were legally entitled to promulgate and which the Indian Company had to obey, and which were of such character as to effectively but legally block the measures otherwise taken or ordered by the parent company to ensure the safety of the affiliate's operations.

A possible issue in this connection arises from the contrasting computerization of certain safety features of the West Virginia plant and the manual operation of corresponding features of the twin plant at Bhopal; was it more than a rationalization to say that the Bhopal plant was being pressured to avoid laborsaving devices given the very high unemployment rate? (And the low wages?) Such allegations are not too difficult to sustain, if true.

What is somewhat more possible is that the Indian company, in conforming to general Indian attitudes and practices, behaved unsafely. But, in this case, Union Carbide U.S.A. would only gain the possibility of suing its own Indian company and officers for bowing to the situation and suing the Indian governments for non-feasance and negligence. It does make sense, and perhaps even good law, to call to account any corporate entities, whether governmental or not, who illegally permit, condone, or engage in corrupt or improper practices that tend to bring about a liability on the part of the multinational company if the company can show, on its side, attempts to resist such behaviors. However, neither the Indian, nor other governments, has gone far in giving up its sovereign right to evade suits.

Union Carbide (USA) may contemplate other arguments, as that it merely offered designs, procedures, and consultation to the Indian company, which could, at will and despite flat orders to the contrary, refuse them. But, in such a case, it would only be setting up a policy, not creating a new kind of entity -- the owned, but legally uncontrolled, subsidiary. Again, if it allowed its powers to go unexercised, Union Carbide would also be proving itself a poor sort of parent, inert and neglectful. And, all too often, the manager who makes no decision when a decision is necessary, may encourage his subordinates to do the same.

We scarcely need to dwell further upon this matter. Union Carbide (USA) controlled Union Carbide (India), albeit fitfully. The former managing director of Union Carbide (India), in a sworn affidavit before a U.S. Federal Court early in February, claimed that he and others in India differed from the U.S. company view on a controversy of the mid-seventies over whether MIC should be stored at Bhopal in small containers or large tanks. The U.S. view prevailed.

Also, a U.S. inspection team was sent to Bhopal in 1982 and reported back detailing a number of practices, including safety procedures, hat required correction and Union Carbide (India) dutifully reported over the next two years its progress in correcting its deficiencies.

Monthly reports were submitted to the parent company. The Indian annual budget was cleared for approval by the parent company. One could bring forward other instances and many more of them must be contained in the records of the two companies, to show that Union Carbide (USA) did govern, well or poorly, Union Carbide (India).

To the fertile mind of a lawyer, a veritable cornucopia of lawsuits can have issued from the killer cloud of Bhopal. Without much ado one counts eighteen parties (or classes of parties) that may be inclined to launch significant types of civil and criminal actions at law. All of these parties might initiate cases against more than one defendant party. I counted as many as seventy such potential parties of the second part. Thus we have an eventual possible total of seventy categories of actions going before the courts. To differing degrees, every resident of Bhopal, Madhya Pradesh and India, and every stockholder of Union Carbide might be involved in the court processes.

Who can sue whom for what? Who can charge whom with doing what? Proceeding to answer these questions, we obtain an acute sense of the far reaching legal entanglements and consequences and meet as well with some surprises. I do not suggest that all of these are viable actions or even desirable; in fact my motive in exposing them is to demonstrate the expanse of the cloud of responsibility that is over Bhopal and to show how interrelated are all the characters of the dramatic tragedy. To exercise all legal possibilities, much less all wrongs that entertain legal hopes, would be a poor solution, though not the worst. The best solution would be a grand imaginative settlement, but I am persuaded that this will not come about without the help of the courts. Indeed the courts are already bringing positive benefits by the mere prospects of their employment.

In the first rank of litigants would come the direct individual victims:

We classify them as the heirs of the deceased (who may number over 15,000, if the dead are 3000, and who may live far from Bhopal); the injured, who may number 200,000; losers of property to the same numbers; and business losers to the number of a million.

Once more it may be well to stress the true nature of the damages inflicted. Putting aside death and personal injury, examine the matter of property losses. A poor child in India may wear a silver bracelet. It is the family's property and will go to her dowry; it represents months or years of savings: its loss should be carefully charged up, not discussed as unworthy of interest in a billion dollar suit.

So with shacks that have had to be sold for a few rupees: these are wrought out of almost nothing, but that is not to say that they, and the site upon which they squat, are obtained by nothing worthwhile. They are homes, they were homes, and their abandonment or sale because of death in the family requires logging in the account books. Everyone in Bhopal lost work time, rich and poor alike, so all of this must be calculated in damages. If a cobbler working in a shop no larger than a packing crate has lost two weeks of work because of personal preoccupations or loss of customers, he deserves compensation. Similarly, the middle class person has lost on a more elegant scale proportionately and requires recompense.

The heirs, the disabled, those undergoing minor injuries but much anguish and suffering, and those with business losses are logically trying to sue the party that is wealthiest, Union Carbide (USA). However, they can, may, and perhaps should sue all four entities, Union Carbide (India), the State of Madhya Pradesh, the City of Bhopal, and the Government of India. Union Carbide (India) can be sued for damages on grounds of negligence, gross negligence and criminal negligence in India, just as its parent company in America can be. The grounds for suit are set forth in the next chapters in some detail. One should not overlook, too, the right of individual victims to file criminal complaints, whether in India or the U.S.A., especially in the event that criminal charges are not being pressed by the authorities where the alleged "crimes" have been committed, Bhopal and Bombay. Further, individual claims may be filed against seemingly culpable top executives and boards of directors. (I have counted neither these nor the preceding criminal actions in calculating above some eighteen opportunities for suit.)

In addition, the victims and damaged -- numbering as high as a million persons -- should be able to sue but almost surely cannot sue the State of Madhya Pradesh, the City of Bhopal and the Government of India, all on the roughly similar grounds of granting licenses without obligatory prior determination of fulfillment of required conditions for doing business, failure to carry out obligatory inspections, and in the event that sworn testimony is available about individual officers, the criminal acceptance of bribes in connection with quick-pass or other inspection failures.

Actually it is conceivable that a case can be brought against the U.S. government before the World Court at the Hague for knowingly and willfully permitting the export of design, factories, management services, and products that are hazardous and pretend to be what they are not such as "equal and uniform" with U.S. standards. Probably, because the U.S. government will refuse to except the World Court, the case will not succeed in a material sense, but the potential, virtual international law of this year tends to become the actual law of the future.

Likewise the City of Bhopal might sue all of the forgoing and perhaps its own parent State of Madhya Pradesh. Fortunately, because the State recently decreed the right of the squatters to their bit of turf, it cannot try to evict all the squatter victims and charge them for unlawful trespass, should it feel driven to commit some spectacular folly.

Let us now proceed to the State of Madhya Pradesh, which can choose civil and criminal actions aimed at six parties: Union Carbide India, of course, then Union Carbide (USA), the Boards of Directors of both companies as individuals, the Plant-Manager, and one or more supervisors and workers. It might even try to sue the Government of India for not doing its statutory duty in restraining and controlling the importation and processing of ultra-hazardous products. The several Union Carbide officers and employees who were arrested were charged under the Indian criminal code with criminal conspiracy, culpable homicide not amounting to murder, causing death by negligence, mischief, mischief in the killing of livestock, making the atmosphere noxious for health, and negligent conduct in respect to poisonous substances.

A month after the disaster, the Chief Minister saw to the suspension of six Labour Department officials. It appears that, following serious phosgene gas leakage accidents in December 1981 and January 1982, an inquiry was ordered. A local science college chemistry professor was allowed to take over two years to report. The report then waited for seven months in the lower offices, and further delays occurred at the level of the Under Secretary and the Deputy Secretary. Only in October 1984 did the Secretary come to know of it. Still no action was taken and when the Chief Minister, a week after the accident, asked for the report, a week passed before he got it. Safety obviously did not have top priority. A seventh official, the Chief Inspector of Factories, was dismissed for having given perfunctory approval to the Union Carbide License year after year "without taking cognizance of the safety lapses." Aside from the accidents, it is notable that of the seven Union Carbide employees arrested following the accident, three had not undergone the training with the parent company which had been assured in obtaining the license setting up the plant.

In entering India with a new process, Union Carbide had to seek a license. Its application proceeded through the Ministry of Chemicals and Fertilizers, the Directorate General of Technological Development, the Ministry of Agriculture, the Central Pesticides Board and several State Government agencies of Madhya Pradesh. As Aristotle pointed out 2300 years ago, everybody's business is nobody's business. The Bhopal tragedy may serve as a guideline for tracing the licensing process in the Indian government. An examination of the pertinent records of all of these agencies (others, too, have had a hand in the matter) should provide recommendations for general reforms of administration. Corruption is prevalent, it is believed, but unlikely here; what is not generally realized in

India is that the murder of time and the dismembering of responsibility may have worse effects than corruption. Corruption, too, is often the only way to dip down into the muck of dismemberment and delay to pluck out a project while it is still alive.

The Government of India might not only sue Union Carbide, Union Carbide India, and the State of Madhya Pradesh (for corruption and nonfeasance in controlling operations at the lethal site), but also it might sue the government of the U.S.A., whether in America, in India, or at the World Court, alleging that the U.S.A. is effectively the incorporating authority, the regulator, the authorizer, the sponsor, and subsidizer of Union Carbide in its operations abroad. The suits are worth pressing, not especially with confidence in their outcome but with interest in their merits. There are significant analogies between the privateers of old, commissioned by a state and let loose upon the high seas to take and destroy hostile vessels (even in the absence of declaration of war) and the irresponsibility with which many nations encouraged the launching of multinational corporations; privateering, it will be recalled, was finally outlawed in international law. Indian environmentalists might mobilize pressures and talents to bring these cases before, if not the courts, then the bars or world opinion.

But then, too, the U.S. government might have a case against the Indian government for offering unlawful and inappropriate inducements to attract a multinational company, and for not keeping its part of the bargain to control and inspect the company. At the same time, the U.S. government should not be discouraged from suing Union Carbide for any possible involvement in improper solicitation of contracts, or conceivably the corruption of local and state officials (who knows what witnesses to this sort of crime may be hovering in the shadows?). When a company and individual acts badly abroad, their nation suffers defamation. It might be nice to think of the U.S. government suing Union Carbide on such grounds, but the idea is fanciful, no more than parents can sue their children for the damage they may do to their parents' reputation.

Now then, Union Carbide (U.S.A.) itself is likely to enable actions against Union Carbide India and its Board of Directors for falsifying records, exceeding its authority, and endangering the interests of its stockholders. Doubtless, too, Union Carbide will be heavily engaged in litigation with its insurers, who will be most reluctant to pay the full amount of insurance coverage called for under their contracts with Union Carbide.

In its turn, Union Carbide (India) must consider whether to sue its parent company for denying it the funds, services, and permissions required to run a safe plant. It, too, will engage in law suits with its insuring companies. And it may even contemplate a suit against the State of Madhya Pradesh for statutory nonfeasance, negligent inspection procedures, and accepting bribery (if any credence can be placed in widely spread rumors).

If, as the two corporations allege, the Indian governmental authorities permitted the vast colonies to be founded near the factory gates, may the authorities be liable in a counter-suit for setting up victims in large numbers for prospective disasters?

The Chief Minister of Madhya Pradesh put his own predicament and defense beautifully when he said, "In a way, I am responsible for everything. But there must be some level at which the persons most concerned have to have greater responsibility than me." Again, it may be difficult or impossible to prosecute such actions in India or Madhya Pradesh or the U.S.A. Still a legal trend may be noticed, which allows suits for nonfeasance and charges of corruption to be brought by victims, even culpable corporate victims.

So, too, might certain individuals try to go to court. The Chairman of the Board of Union Carbide (India) may elect to sue his own company, Union Carbide (USA) and the State Government with somewhat the same rationale:

failure to provide information, non-compliance with originally agreed conditions of engagement, false arrest, non-feasance, and corruption.

The same logic, more forcibly perhaps, may be brought to bear on behalf of the Plant Manager, who appears to be the most vulnerable official of all, and in regard to whom high technical managers around the world will attend with some sense of identification. Can a Plant Manager, hired for his technical engineering and management abilities, sue his company, his parent company, and the State of Madhya Pradesh for denying him the equipment and personnel needed to ensure safety at the plant site, refusing him necessary funds for this purpose, neglecting to give him the back-up support of information and inspection he needs, also for false arrest, malfeasance in failure to inspect required by law, and solicitation of bribes.

There is a Union at Union Carbide (India) that claims the company has never responded to complaints about its safety. Its members have lost their jobs (though still paid) and will find it difficult to get new work; many will have to move out of town and will suffer from the reputation of the disastrous plant. Its suits lie against the State as well, if it can show that its demands for inspection and compliance to the law were ignored or deviously avoided. There was a lot of union-bursting going on at Bhopal; is there a tort here, if only on the safety question, that could be pursued in the American Courts?

If the multinationals were to organize around the world to regulate themselves, they might as well seek the legal power upon entering a country to sue the governments. Just as the governments can hold them accountable for a breach of agreement, they might hold the government accountable for the same. There is perhaps no good reason, whatever the legal situation, why the Chief Minister should be able to say that the burden rested with the Company to inform the local authority about potential hazards, but then the Company should not be able to say the same of the local authority. If a company contracts with a private security company for guarding its property and the private police wink at thefts, let the plant be flooded, and disconnect the alarm systems, a suit for damages would legitimately ensue; why then should a government be able to violate its own explicit and implied promises of inspection, zoning, public health measures, and fire and police protection?

The shareholders of both Union Carbide (USA) and Union Carbide (India) may request their day in court; the first against their own Company, the Government of India, and the State for reasons already given, the second against their own Company, Union Carbide (USA), and their Board of Directors. For instance they might sue Madhya Pradesh for nonfeasance under a 36-year-old Factories Act that directs States to make rules governing the proper site surveys and the legitimacy of various manufacturing processes prior to permitting construction. Such rules have never been promulgated by the State. There are other improprieties, too. But again the "sovereignty" of a government mocks promises and injuries.

The hospitals of Bhopal and the Medical School which performed nobly and heroically during the tragic days of massive admissions are poor and underequipped and ought certainly to present a bill for services rendered to the Union Carbide Companies. Since some number in the order of 600,000 stays or visits were accomplished, and the most modest American bill would average $50 per treatment, 30 millions or some lesser compromise would be a reasonable charge, or, if payment were not promptly forthcoming, might be the amount demanded in a case at law. (Of course, a great many patients received inadequate, mistaken, delayed treatment by doctors and hospitals, for which, in America, under normal conditions they could sue; what kind of case law applies to mass emergencies? Hospitals had better seek appropriate legislation.)

A number of insurance claims will undoubtedly enter court. The principle behind insurance against damage claims brought by employees and injured third parties, whether individuals or private and public entities, is that a heavy realized liability of a single insured can be spread over a great many insurance-holders. At Bhopal, it is apparent that the Indian insurance held by the Company, guessed to be $200 millions, is altogether inadequate to pay off even some unlikely moderate settlement or judgement. Moreover, it is improbable that such insurance would be paid all or in part by the primary insurance company, given the negligence aspects of the case.

The situation is similar and still more complicated with regard to Union Carbide (USA), which has an immense umbrella composed of several agents and insurers. Should Union Carbide be assessed for punitive damages, it will be in a most difficult position, because even if any insurance resources were yet untapped and even if the insurers had contracted to pay for punitive damages, the U.S. courts have not permitted this to be done in times past, holding that if punitive damages could be covered by insurance then their rationale as a punishment would be largely defeated.

Punitive damages are intended to punish a company that recklessly disregards safety in the making or handling of its product. An intentional, deliberate choice of the non-safe or risky over safe materials and procedures gives reason for punitive damages. Thus punitive damages go far beyond the calculated needs of the victim and makes the victim beneficiary of public policy in a civil case, where jail terms and heavy fines are not at issue. Triple damages are common, but larger multiples are not unknown.

The question of whether any award should be so large as to destroy a company is, strictly speaking, irrelevant to a court's judgement. In the Manville Corporation asbestos exposure cases, indications in many pending cases that the company would be found to have knowingly and deliberately exposed its workers to risk of severe pulmonary and other disabilities led to the voluntary bankruptcy of the corporation. Union Carbide must be considering the same way out of its jungle of problems, and may even regard it as one of its trump cards. For, if its Board of Directors should decide to prefer certain creditors to others and to engage in certain personally or even publicly beneficial maneuvers under the umbrella of bankruptcy, they may deliberately plunge the company into a voluntary bankruptcy, abandoning their stockholders for the larger part and letting the legal process and relief of the Bhopal victims extend almost indefinitely into the future. This could be a second Bhopal tragedy.

One can foresee costly litigation extended over some years. As with the others types of litigation arising out of the tragedy, the insurance litigation will be legally interacting with the litigation of other parties in other courts, until some key determination is made somewhere, that begins an avalanche of settlements and judgements. To predict the champion in this mass marathon to resolve legally the Bhopal tragedy is almost impossible.

One of the more likely resolutions would be the collapse of Union Carbide Corporation under the weight of financial threat, followed by a court dismemberment of the company and parcelling out of its assets to innumerable approved claimants. (It is likely but not certain in law and in the politics of the collapse that the victims would receive priority in the distribution of the proceeds.) One should appreciate the size of such a transaction: Union Carbide has sales rivaling the combined gross domestic product of Zaire, Zambia, and Zimbabwe; imagine the scene should some mythical world authority decide to dismantle and reorganize them because they failed to pay their foreign debts.

One may profit, though, from the analogy to afford some sympathy to the hundreds of thousands of stockholders, suppliers, employees, and officers of the Union Carbide complex. A majority of these will be adversely affected, whatever the resolution. The Chairman of Union Carbide was prompt to say that his life would never be the same again owing to the tragedy. The mood of the people at Bhopal's sister plant in West Virginia was of general shock and mourning. Granted that all will have to swallow their feelings and losses, and submit to tinges of regret and guilt, still they all, too, will have a stake, as tiny or large as their sympathies, in a human and creative resolution. Indeed, with the Bhopal disaster, Americans on the whole must add another iota of guilt to what has been accumulating since the decimation of the U.S. Indian tribes, the suppression of their black fellow countrymen, the extirpation of German cities, the holocausts of Hiroshima and Nagasaki, and the ravaging of Vietnam. Therefore, the American people, too, have an investment in the fate of the company gone abroad, for they have become party to its policies and disasters.

CHAPTER VI

Causes and Negligence

The day after the killer cloud struck Bhopal, Union Carbide officials at their Danbury headquarters were telling the press and public that they would require two to three weeks to determine and report upon the cause of the disaster. Ten weeks later they had provided no account of it and were carefully maintaining secrecy. Then again they promised the report in two or three weeks.

Furthermore, the Chairman of the Board said: "Our report is restricted to what happened in those tanks. Our report will not deal with how things were done in India from a managerial or a personal standpoint. I don't think you have to reconstruct what the people in India did." This is an enunciation from the foxhole of legal defense. It tells us to expect a technical report of limited scope.

In fact, the first news of the Union Carbide report reaching India on March 21 indicates that it is the kind of technical report whose premises are concealed -- intending thereby to focus blame upon purely technical negligence and personal negligence at the Bhopal site. It is presented as the culmination of a massive research project. It hints, moreover, at the possibility of sabotage, a deliberate act in releasing water into Tank 610. It appears to reflect on-the-spot investigation after the accident; promptly government officials at Bhopal denied that the American team had been allowed entry into the pesticides unit, denied that the Indian Central Bureau of Investigations had given them any access to relevant materials, and asserted that the Union Carbide experts were not even allowed to speak to their Indian workers.

My aim here is of course broader: "to reconstruct what the people in India did" and to "deal with how things were done in India (and America) from a managerial or a personal standpoint." Probably this report will be more helpful, even for the Company.

The ultimate cause of pesticide accidents is the pest. Pests destroy crops, spread disease, and convey endless annoyances. They proliferate; under pesticide pressure, they also may change genetically to defend themselves. Means of combating pests are numerous and the most effective of them have been toxins in a form for wide dissemination, poisonous to humans as well as to pests.

Invention is a human non-genetic change; pesticides are continuously in the process of invention. As with all inventions, a time lag occurs before the adoption of new types of pesticides and techniques. The lag is brought on by dislike of losing one's investments in old techniques as well as by the time required for rendering an invention practical on a large scale.

Perhaps this needs be said because in the Bhopal case one perceives both the pressure of new pesticidal inventions and equipment and the resistance to scrapping investments in old formulas, plants, and procedures. These, too, may be termed causes of the tragedy.

As one works closer to the tragedy from such remote causes, he comes upon many a closer cause, so many that practically everyone whose behavior is mentioned in this report can, whether he wishes so or not, be placed in the network of causes with some justification. For legal and journalistic purposes, a disproportionate amount of attention is invariably to be given to the immediate cause, the "trigger-man," whoever committed the "one" act without which no gas would have escaped and the City of Bhopal would have rested in peace during the night of December 2-3, 1985.

Such a "trigger-man" appears to exist. He would be the worker who fitted a water hose into a pipe that sent or leaked water into Tank 610. What the American Chairman of Union Carbide implied of himself after the accident is true of this worker as well: our problem is not solved by nailing him to the wall. However, for purposes of delineating the chain of causation, it is as well to begin with the worker. He is known to authorities and Union Carbide and has been mentioned in the press.

The complex of buildings, machines, pipes, and toxic chemicals in which our supposed "trigger-man" was to be found working that night may best be conveyed by a set of diagrams. These show the factory layout and neighboring area (Figure 2). The MIC storage tank (Figure 3) and the vent gas scrubber (Figure 4). The path of the gas cloud vented above the scrubber has already been shown (Figure 1) as it affected the City of Bhopal.

A stainless steel tank emplaced in concrete contained probably 45 tons of liquid MIC. The Union Carbide *Manual* calls MIC ($CH_3N = C = O$) "an extremely hazardous chemical.. by all means of contact" and regards it "as an oral and contact poison" even though it is not classified among poisons. It is also "extremely flammable." Most probably, water got into the tank through a pipe. The MIC, which reacts violently with water, turned into an explosive gas vapor that blew out the valves in its path.

The event was a constrained explosion, not a leak, and the explosion formed a cloud which blew downwind over Bhopal. It was the simplest of occurrences: a tank of volatile liquid, a violent reaction with water, a prolonged explosion of gas through a pipe and out.

Sometime after 9:00 P.M., with the night shift due at 10:45 and not much going on, the "trigger-man" was taking a cup of tea at the company canteen. He had worked for seven years at the plant, and for reasons unknown, two months before, had been transferred into the unit that makes and stores MIC. He had less than the background and training originally required to fill his job as an operator.

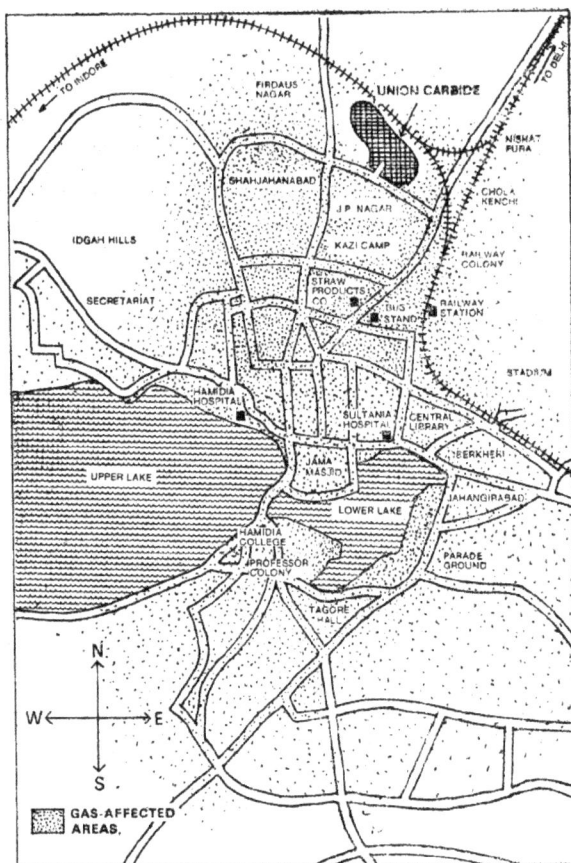

CITY OF BHOPAL SHOWING GAS AFFECTED AREAS

Figure No. 1

His supervisor telephoned him from the MIC area office to come over and clean a pipe. Cleaning the pipe in question was on the job list for the shift, possibly the only notable chore on the list. The supervisor himself had also joined the MIC unit two months before; his prior experience had been with the Union Carbide battery division in Calcutta. He met the operator at the MIC area and showed him a 25 foot pipe, 8 feet off the ground, that was to be washed. The pipe is said to have connected an MIC outlet from the manufacturing plant to the MIC storage tanks. The operator proceeded to open a nozzle on it, affix a water hose and flush out the inside. A drain hole in the pipe was opened to let out waste water, which then flowed onto the floor and out a floor drain. The supervisor stood by. Then the two men left the area. The water was running.

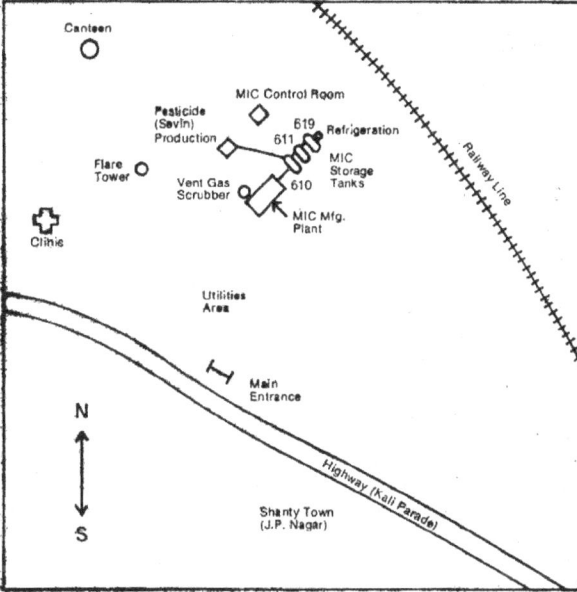

THE MIC SETUP AT BHOPAL UC
(NOT DRAWN TO SCALE, NON-RELEVANT STRUCTURES OMITTED)
Figure No. 2

Figure No. 3   MIC STORAGE TANK

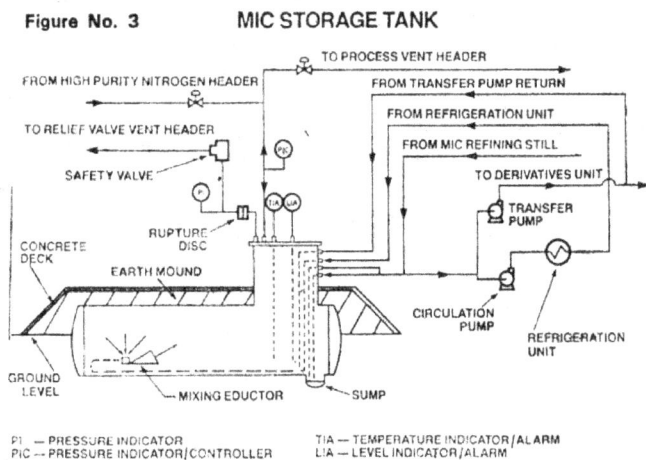

I do not know the flow rate; it would have been heavy; in the three hours that is was running, I assume that enough water flowed to fill a large tank even while draining. Nor do I know the amount of water used ordinarily to flush a pipe of MIC residue. Nor is there any indication that either man checked to be sure that the pipe valves that blocked passage to the tanks were all closed. Could they possibly have thought that they were to clean out the "empty" tank 619? Not if the work agenda said to clean out the particular pipe. Is it conceivable that one or the other in a moment of confusion mixed up Tank 610 with Tank 619 and actually began to flush out the full tank 610? Probably not, although why so much water should have been used and let run on and on -- and forgetfully as we shall see -- is still incomprehensible.

Now, more negligence and another odd fact: the overfull tank 610 had been filled around October 22, for the MIC plant had been shut down since then. That is 40 days from December 3, over twice the recommended storage time limit. This fact is damaging enough; yet it lets us also think that these men may even have forgotten what Tank 610 contained! And we know and will know more about the respect accorded records in this setting.

Thus far, the "trigger-man" is just that and acting as a dutiful worker. He noticed that the valve leading to the MIC tanks had been sealed. (It must have been the valve to Tank 610, or that the valve to Tank 610 was open enough to receive leaked water from the first valve through a pipe common to all three storage tanks).

Then the "trigger-man" may have joined the ranks of the culpable. For he noticed that the closed valve had not been sealed by the extra metal disc, "a slip blind," that is standard and required operating procedure when contaminating substances are present in a pipe leading to any MIC storage. And he did not insert the slip blind. Now did he tell the supervisor to do so. Nor did the supervisor act on his own initiative. Or so it appears. And the truth will not be fully known until both men and several other workers testify to the events in a court of law where liable to perjury charges, if then.

The "trigger-man" asserted to Stuart Diamond of the *New York Times* that "it was not my job," those fatal words so painful to the ears of managers, leaders, and theologians. It would be psychologically of interest to probe further how this utterance emerges from this particular man, interrupted while drinking tea by an inexperienced new supervisor, and called to a chore shortly before a change of shift, himself poorly trained and unafraid of MIC except as an irritant (he said), himself new to this unit (was it a demotion, was he being paid less than the job specified?) and having heard the common expression that "valves around here are leaking all the time," and being party to a general decline in morale at the plant. Cause after cause: perhaps something so small as a supervisor who may not have ordered a subordinate about in the right tone of voice.

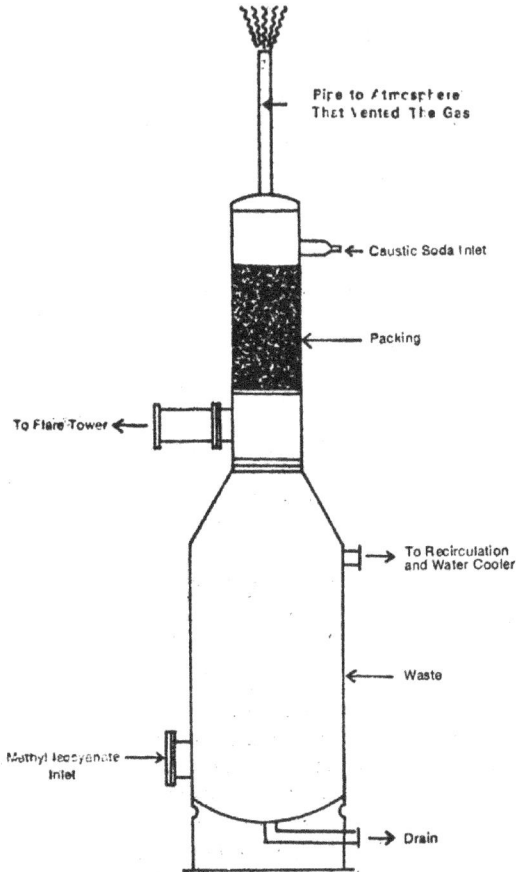

VENT GAS SCRUBBER
Figure No. 4

In any event, by 9 : 45 P.M. the water was running.

As the work shift prepared to retire, almost an hour later, the gauge in the control room for Tank 610 registered two pounds per square inch (2 psi) and was so logged. This was considered normal tank pressure. Temperatures were not recorded in practice; so this critical measure was and is unavailable. Ordinarily the temperature of the tanks hovered around 20 deg.C (68 deg.F), well above the recommended storage temperature of 5 deg.C (41 deg.F). (The Union Carbide *Manual* says: "Maintain a tank's temperature below 15 deg.C (about 60 deg.F) and preferably at about 0 deg.C (32 deg.F). Equip a storage tank with dual temperature indicators that will sound an alarm and flash warning lights if the temperature of the stored material rises abnormally."

The night shift came on at 10:45 P.M. At 10:00 hours, the control room operator noticed that the pressure on Tank 610 registered 10 psi, up by a factor of five from the 2 psi logged earlier.

The supervisor noted the rise also a half-hour later but thought the original reading might have been faulty. The men also thought that the pressure might be up as part of an expected pressure rise occurring in a sister tank, 611, where they believed that nitrogen gas was being used to push the MIC liquid into the mixer in preparation for the manufacture of pesticide. (This implies that they regarded such a "backlash" as not uncommon.) Now it was 11:30.

Between 11:30 and 12:00 employees in the utility area sensed a gaseous irritation of their eyes. They regarded it as some tiny leak, not rare. However, by midnight the leak is sensed generally around the MIC unit and the supervisor is told the news.

The workers begin to look around for leaks. One of them spots a drip high on the MIC plant wall with a whitish gas exuding from it. He reports it to the night shift supervisor, who may believe it to be water, and the supervisor is reported to say that he would examine it after the 12:15 tea break. This leak would seem to have been a true MIC condensation and spume, backed up through the pipe being purged and through the leaky valve, which by now would have had a heavy reverse pressure moving against it from the gas in Tank 610.

The operator in the MIC control room now reports that the pressure and temperature in Tank 610 has risen much higher. The supervisor and an operator go over to

Tank 610 and find that the rupture disc has exceeded its bursting point and has blown; in addition, the safety valve backing up the rupture disc has popped out. The temperature is well over 39 deg.C (102.4 deg.F, the vapor point of MIC) and pressure well over 40 psi.

Thinking now desperately of what might be the source of the trouble, they turn off the water washing the tubes. It is about 12:45 A.M. a few seconds or minutes later, a vapor plume emerges from the nozzle of the vent pipe some yards away and 100 feet up, and the men know that an accidental discharge is occurring and can only hope that it is limited and somehow will lose itself high in the air. A worker turned on the scrubber which was intended to neutralize the gas with a caustic soda solution. A loud alarm was sounded for several minutes to warn the public beyond the factory walls, followed by and reduced to a muted alarm to warn the factory workers.

By this time, too, large cracks had been seen and heard to appear in the heavy concrete housing of the almost totally encased MIC tanks, and the prospect of an imminent explosion of the tank itself must have entered the workers' minds. Several had donned gas masks. Water was poured on the tanks. Hydrants and hoses could not reach up to the gas cloud itself, where it was escaping from the vent pipe. There seemed to be nothing to turn on or off that would help. Most fled upwind. The supervisor's gas mask was lost or used by someone else; he was gassed and then injured in climbing over the fence. No one else was hurt.

Around 1:00 A.M., through a patrolling inspector, the city police learned that something serious was up. The police control room was affected by the gas, and telephoned Union Carbide only to be told that nothing of importance was occurring. The Police Chief had two men trying to call the plant. Between 1:25 and 2:10, he got through three times. Twice he was told: everything is O.K.;

then, that they didn't know what was happening. This was reported by Praful Bidwai in *The Times of India*.

A district magistrate heard of the action and called the Union Carbide Plant Manager, at 1:45 A.M. The Manager drove to the plant sensing the gaseous air *en route*, but could do nothing upon arrival. He himself thought for a while that the leak had been plugged, and said so.

Ever since the disaster, other ideas have contended with the leaky valve theory. One argues that metallic or other substances might have contaminated the liquid of Tank 610. Tests have now discovered traces of iron, lye, water, and phosgene in the empty tank. The head government scientist in charge of the conversion of the remaining MIC to pesticide in "Operation Faith" pointed out that phosgene, when combined with water would produce chloride ions that would corrode the steel of the tank, producing iron, which would accelerate a reaction with the MIC. However, the trace measures in parts per million are not yet available, and nothing argues for the presence of more than the smallest amount of iron, and all phosgene other than a minuscule fraction to stabilize the MIC is denied. Anyhow it is doubtful that a tank which had been standing quiet but whose pressure rose from normal (2 psi) to a valve-bursting point of above 40 psi in two hours would have done so on the basis of a water/iron-phosgene-chloride-iron-MIC sequence in so brief a time, especially since the phosgene was probably available in only minute amounts. It is more reasonable to theorize that the iron and chloride ions were generated almost wholly in those fatal couple of hours.

As for the lye, *The Hindustan Times* heard initially that workers had used lye in a cleanup of the area around Tank 610 the day before, but there is no evidence that lye entered the tank and in any case the reaction time would have taken too long and the gas created would have been too voluminous. The surmise is hardly tenable.

Another idea is that Tank 610 contained some new chemical diabolically conceived in the Research Laboratory and this substance either went out of control or was used to experiment on the poor people of the area. Naturally the CIA has been mentioned in this connection, and I would suppose that this suggestion ought to be ignored or else become the subject of a legislative or other public hearing in which one might at least put the allegation to rest and discover who are the public paranoiacs in the setting.

A favored idea among the chemical cognoscenti is that MIC might begin to transform from simple to giant molecules, that is, to polymerize, under favorable conditions of heat and temperature. Perhaps the chief obstacle to this idea's progressing is that water was definitely in the picture in quantity, and that the time elapsed between a normal and an explosive temperature was too short for the process to occur. Such molecules were reported discovered in the analysis of the tank residue, but these were probably formed under the heat and pressure of the total reaction.

I have already mentioned the incredible idea that the blunder made was stupendous, that the supervisor and operator somehow confused Tank 610 with "empty" Tank 619 and thought they should wash out "empty" tank 610. But then several items contradict this: the pipe drain was allegedly opened, the MIC valve was allegedly closed, the water was running so long accidentally, and so on. If any item makes this notion plausible, it is the long three hours that the hose was running, and the large amount of water that got mixed up with Tank 610. Otherwise, the idea has small merit.

But an estimate that at least 1.5 tons (450 gallons) of water was needed to gasify 45 tons (13,000 gallons) of MIC, is troubling. Few can believe that the valve in question could be so faulty. Still, the water would have entered the liquid from above. The main reaction would be occurring near the top of the overfull tank. Shortly after the rise in pressure, the newly formed gas might bring pressure against the leaky valve -- not enough to push back the water leak, itself being hosed under pressure, but enough to create a path in reverse through the valve. Somehow, we recall, gas backed into a pipe and was emitted and condensed high on the wall of the MIC plant.

The reverse gas pressure would strain the faulty valve. The valve would admit water at a faster rate. Soon it would be pouring at a rate easily capable of supplying the necessary 450 gallons within an hour's time. A concentration of MIC gas beyond all experiment and experience was primed for release.

Let us now put ourselves in the position of those designing a plant to make MIC and to convert it into the powdered pesticide, Sevin. We are thinking now of industrial safety, having already settled upon the manufacturing process. First of all, we would ask the nature of the process and the risks attendant thereto. The parameters of risk at all stages of handling and of all types of mishap should ideally be known and incorporated. Presumably no material would be handled nor any process conducted whose capacity to harm was uncontainable at a foreseeable level of probability of intensity of occurrence.

We would wish a site removed from settled areas to limit the effects of any accidental release of our noxious products, and seek guarantees of zoning against the near approach of any residential population. We would locate the more dangerous products and processes away from the rest and sheath or implant them securely. A central control room would be set up away from on-the-spot controls, and provide for around-the-clock control procedures for inventories and input-output processes, such that gauges, human senses, and automatic signals doubly or trebbly reinforced, would collectively monitor them. Any exceptional or abnormal procedure or process, whether deliberately or accidentally undertaken, would demand attention and a double "go-ahead" approval, involving upping of level of command. Full gear would be provided for risky routine operations, and some person would be charged to record all uses of the gear.

There would be an imperative automatic and human response to any emergency with risk of fire, explosion, or contamination. Access to all points of risk would be assured either by automatic equipment or protected human intervention. Back-up large-scale emergency equipment for fire, gas explosion, and injury could be set up.

Specially invented state-of-the-art devices would be activated automatically or humanly at points of high risk, with levels of activation scaled to the level of possible emergency challenge. Secondary reinforced systems of emergency technical assistance and first aid would be provided by cooperation with neighboring public agencies -- fire-fighting, hospitals, ambulances, security and police. The effectiveness and availability under all circumstances of state-of-the-art telecommunication would be assured.

Although the foregoing physical design objectives would be adjusted to and reinforce normal human capabilities, the human behavior corresponding to the physical and mechanical capabilities of the plant would relate to them like one map overlaying another, such that a point of risk is overlain by reinforced human behavior just as it is by reinforced physical safeguards (this is the principle of "redundancy.") The safety of every point of risk should be backed-up by the availability of replacement parts, to minimize time of exposure to higher risk. No machine-human system should be able to do what it should not do and every machine-human system must do what it should do. Machines and humans should interact with process and material in such a manner that safety is assured at the highest level of probability.

Listening to and reading what all concerned have said about how matters were conducted in the Union Carbide plant at Bhopal and reading what was written as to how things and people were to be managed, it would appear that at some time in the past every eventuality was thought of and every contingency was provided for.

Let us match this brief and abstract prescription with the realities of the Bhopal event, taking the human and physical together, as one should, in the interacting human-material system of work at the plant.

No one, it seems, from the first-level operator up to the Chairman of the Board of Union Carbide USA knew or admits to having known of the mass-killing potential of MIC under industrial conditions. This is a top management failure, a scientific failure, a judgemental failure, an exorbitant assumption about risk, and a failure of legislative and rule-making bodies.

Indications are strong, too, that, whatever the failure to understand potential risk, a double standard was tolerated, if not applied, by Union Carbide (USA) with regard to U.S. and Indian factories. This concerns personnel, supervision, inspection, salaries, control instruments, and general concern. The intervening structure of an Indian corporation could be used to rationalize the double standard for almost every kind of neglect, it seems, except for the bottom line, profitability, "I do not have a detailed understanding of the Bhopal facility," said the Union Carbide's chief of environmental affairs and safety, who, the *New York Times* reports with nice irony, "has been the company's chief spokesman during the crisis." Further, there had been no U.S. supervision in a year, no work audit in over two years.

Union Carbide needed its broad acreage and could find it on the outskirts of Bhopal. Habitations already existed thereabouts, and by the inexorable law of population mobility, thousands of the poor people settled nearby even before the plant was finished.

Although such a law has been proposed, Indian law does not yet forbid chemical plants in urban places. However, permissions for such construction could have been denied at any time.

The Bhopal city plan and its zoning requirements were violated both by the city and by Union Carbide (India), with mutual awareness. Occasional warnings of danger were ignored. The costs of moving the plant, a million dollars, could have been paid for by the increased value realized on a sale of the land; the idea of moving was conceived solely as a cost item by the government and they were reluctant to impose it on the company.

The plant designs appear not to have worsened safety problems. Further study by industrial architects and designers is in order. The control room, a separate installation, may have been unnecessarily close to the sources of danger, and insufficiently insulated: it filled with gas during the crisis, and blocked visibility as well. Most of the 75 employees on duty were outside the MIC area, and had no trouble in escaping upon warning; they might well have been hurt or killed if the tank itself had blown up.

Pipe and valve identification depends upon their logical, physical markings, the absence of unnecessary visual confusion, and the careful instruction of employees. All three were deficient at Bhopal, as the 1982 inspection plus post-disaster comment show.

The tanks presented several safety problems. They were unconscionably large to begin with. Our first principle -- know the extremity of the risk -- was violated. The Indian engineers opposed the tanks initially, preferring steel drums. Drums require more human attention, a larger staff. It became contradictory to insist upon tanks and then for a chief engineer and inspector of Union Carbide (USA) to say afterwards that various automated safety devices of the U.S. plant were not used in India because the Indians preferred to employ more people.

The stainless steel tanks were imbedded partly below and partly above ground and sheathed with six-inch concrete, leaving a manhole cover with gauges and pipes together with other pipe leads. One may grasp why a generator of potentially dangerous heat should be insulated from the outside air to protect it from external heat and insulated to refrigerate it, but then why should the outside sheath be in effect a concrete bomb casing? Would the water sprayed upon the concrete in an emergency reach and affect the interior heat or would it be less effective?

In the U.S., underground tanks of hazardous chemicals are not controlled by legislation or rules of the federal government, but by local laws, chemical associations, engineering rules, and the ever-threatening potential of personal injury and property damage. Tank regulation in India is almost non-existent.

The situation further suggests that Tank 610 had been overfilled by a batch of MIC from the plant and then the workers decided that they had better get rid of the remainder by storing it in the tank that they knew should remain empty. Workers have asserted that the 15,000 gallon tank contained 13,000 gallons (45 tons, 87% capacity) while the Company had said 11,000 (73% of capacity) and the company rules recommended 60% or, elsewhere, better, 50%. The Manual on MIC can be quoted:

"For safety reasons, size the tanks twice the volume required for storage. Use the added volume, in an emergency, for space to add inert diluent as a heat sink; addition of a diluent will not stop a reaction but will provide more time to control the problem. As an alternative, keep an empty tank available at all times. If the methyl isocyanate tank becomes contaminated or fails, transfer part or all the contents to the empty standby tank..."

Significantly, workers said that Tank 610 had refused to take nitrogen pressurizing the week before; it is by introducing nitrogen into the unfilled portion of a tank that of MIC into the pipes, proceeding to the pesticide (SEVIN) manufacturing plant. Again a disturbing thought occurs. If the Tank had been overfilled to 73% of capacity, as the Company says, and overfilled to 87% of capacity, may it not be possible that the Tank refused to take nitrogen pressure because it was filled at 100% of capacity?

It is astonishing how much trouble three tanks and a few drums can cause. The word used to describe the MIC unit before the accident was "inoperative", and the MIC of the tanks as "held in storage;" but there is no such state as "storage" for MIC; it is always ready to react. The volume of liquid in two tanks was unknown, before and after the accident. The existence of five drums of 1.5 tons total of MIC was unknown until the time came to dispose of them. Not only was the level of Tank 610 unknown, but tank 619 was supposed to be empty: then its level registered 3,300 gallons; then 437 gallons were removed to empty it. (We reserve the possibility that water or hot vapor from 610 may have leaked to 619 from the running hose or growing pressure and brought on an escape of the difference in the two measures. That would have brought the total release to about 48 tons.)

That 619 contained MIC violates the company's rule to hold one tank empty against an emergency during which the contents of a leaking tank can be shifted to a different tank. The workers in the emergency thought to open the pipes between the two tanks but refused the option because they did not want to chance another disastrous reaction in 619. They perhaps did not know, too, whether 619 held nitrogen or atmospheric gas.

But if the storage tanks were designed to cause trouble, the gauges and valves of the MIC unit were and had been tempters of disaster. The Union Carbide Manual says that "the pressure in the tank will rise rapidly if methyl-isocyanate is contaminated. This reaction may begin slowly, especially if there is no agitation, but it will become violent." But if workers do not believe the readings of their gauges and, worse, if they are as often as not correct in their belief, the first indication of serious trouble will be irritated eyes viewing wisps of vapor. The Manual enjoins its readers to use their instruments, not their tear ducts, to watch for gas, but such was not Bhopal practice.

The pressure gauge of Tank 610 in the control room passed from 2 psi to 10 psi in a few minutes. Only rank incompetence can explain passing over a five-fold pressure increase in Tank 610. Despite the Manual's vague warning, the notion of "exponentialism" was not part of the training and, perhaps with the decline in educational background, not part of the mentality of the personnel.

For that matter, there is no evidence whatsoever that the higher echelons had any notion of the educational creed of teaching continuously all one can to those below one -- a cultural failure often enough encountered in America and especially serious in the Third World where authoritarianism, rote-learning, and formal course work eventuate in heavy blockages of learning afterwards on the job and in life.

Record-keeping on inventories and operations appears to have been in a deplorable state. Of the three tanks, only Tank 611, in use then, had a known level of 15 tons. Valves were not regularly inspected and repaired and replaced when defective; nor were gauges regularly repaired; there seems to have been little of the systematic scheduling of tasks befitting a largish enterprise employing more or less a thousand workers in a risky environment.

At the present stage of automation, it should have been possible to operate in the control room a twenty-four hour process-imaging diagram, replete with indicators of the state of the total plant and equipped with signaling devices that would call specifically and imperatively for attention to any unwelcome deviations from the norm. (Indian design capabilities are quite adequate; it is noteworthy that Indian businessmen are largely neglectful and ignorant of the large reservoir of imaginative scientific talent in the country, possibly because so many of them come from a background of traders and merchants.)

Company manuals stressed the need to refrigerate the liquid MIC, saying explicitly that any reaction owing to contamination would then be slowed considerably, allowing more time to bring in help and to control it. Yet the refrigeration system through which the liquid would pass for cooling had been shut down for weeks. Even the Freon used in it had been removed for use elsewhere. Instead of storage at 32 deg.F, the liquid was maintained at about 68 deg.F, a temperature exceeding what a company manual claims to be safe. In the event of a runaway reaction, however, the system would have soon collapsed and burst at all its joints.

One tall structure on site is a flare tower. When operative, gases that cannot enter the scrubber or are not dispatched by the scrubber are piped to the tower. The top is lit, and burns off any extruded gases. Theoretically, the tower might have burned off several tons of vapor. (The pressure actually might have extinguished the flame.) Like the refrigerator, it was out of action. A length of corroded pipe had been removed and not replaced.

The operators turned sprays upon the tank. A fire brigade appeared quickly and turned on more water. The tank reaction was too far gone to be more than negligibly cooled. Nor could any water source reach to the gas cloud over 100 feet above nor was there a high-pressure pump to help out.

The pressure gauge of the control room for Tank 610 rose quickly to reach and surpass its maximum reading of 35.55 psi. The gas pressure exceeded 40 psi bursting past the rupture disc and valve leading to the vent gas scrubber. When it works properly, the machine (see Figure 4) admits MIC vapor into its cylindrical chamber and mixes into it a solution of caustic soda. The gas collapses into slurried residue and can be piped off in due course.

Gaseous wastes, including any MIC that evades the system, would be vented 100 feet in the air. The venting was never expected to be large enough in volume to threaten the health or anyone. Since MIC was not being deliberately processed, the scrubber was not operating. The scrubber was turned on manually at the first sight of the plume whistling out of the vent pipe. However, it seems that hardly any gas was scrubbed. A Union Carbide Vice-President, before he resorted to silence, was reported to say that "the neutralizing process requires a certain amount of residence time for the gas to be run through all the scrubber operations. The time was just not there. Where gas gets under high pressure, it escapes at high velocity." As to why valves broke instead of slowing the flow, intense pressure was blamed.

However, some say, the caustic soda solution was weak and additional soda was not fed in during the excitement or was not stored and available to feed in; others say, the scrubber never reached full effectiveness; still other argue that the soda pump was dismantled for repair.

With the delays and the weak solutions, only several tons of MIC, if that much, would have been detoxified. At this writing it is not known if an analysis of residues at the base of the scrubber vat has been made, and if so, how much activity is indicated by it. What seems certain is that the scrubber was not designed to cope with an emergency discharge, of such proportions, even if the system were in perfect working order. In any event, most of the gas would have found its way up and out the vent gas stack. Moreover, the scrubber's maximum operating temperature was 150 deg.F while the temperature of the exploding gas was estimated to achieve 400 deg.F. Finally, a valve connected the scrubber to its gas vent. To have shut this valve, which could only be done manually and at a considerable height, might have killed the hero attempting the job, and might have been succeeded by a bursting of the valve, or cracking of the scrubber bottle mechanisms, or both. In sum, the scrubber was not designed to handle the emergency, it could never have done so, and anyhow was not put to the test.

Gas masks of short duration were available and were donned by some workers as they scrambled about attempting the impossible. They were then cast aside. Buses were available for escape and by some incredible scheme were to have the double function of alerting and evacuating the neighborhood shanties. They remained parked. The fleeing employees trusted more to their feet. A loud speaker worked, and hurried the employees along. A medical facility stood near the northwest boundary of the plant side but went unmanned and unused. I mentioned earlier the sirens, and the early misuse and the belated resounding of the public siren.

The plant superintendent came hurrying over, but, helpless, left with the rest. Diamond, who inquired, found only a lone worker who had manned his post in the control room through it all, staying until the next afternoon; why? -- because there was nothing else to do. That the plant manager was not informed until a magistrate telephoned him says something about the state of awareness and training. Somewhere on the premises there existed a short-range radio for alerting managers at their homes; or so it is said.

Plant personnel, from top to bottom, had improper or inadequate knowledge of a need to communicate in-plant emergencies to the affected outside world. The police were not informed and may even have been misinformed, in either case deliberately as a matter of company policy (for "small" accidents, of course, but "reassurance" has no bounds). Inspections were avoided and underplayed as a matter of course. The public siren may have been deliberately shut off after its first (mistaken?) blast, and only re-sounded much later; so as not to alarm people, they say. These several episodes point to an evasive uncooperative relationship with the public and authorities. I should say, the authorities who had an operational role in connection with the plant; earlier I mentioned the comforts and hospitality extended to politicians.

We return to the "trigger episode": only the failure to slip into place a metal disc to block water from reaching beyond the pipe being washed produced the Bhopal disaster. The many deficiencies of design, management, competence, and safety procedures in this plant would go on producing more and more accidents, perhaps even a large-scale one. But this particular disaster would not have occurred.

When, some days after the accident, it was decided that the remaining MIC must be processed to get rid of its danger, a quarter of the population again fled the city, despite what were termed "additional safety measures." Actually, these were almost useless, except to prevent even greater panic: Union Carbide engineers of the U.S. and India operated the plant, while a flimsy cloth fence, some wet blankets over the gas vent, stand-by water hoses, and a helicopter with water to spray the "cloud" were arranged and called safety systems. The city was controlled as an armed camp on the alert. The operation went off without a hitch. Like so many other industrial and group operations, MIC can be handled safely and has been, if only -- if only -- the operators are mindful and know what they are doing. Still, what is put so simply can be impossible in practice, depending as it does upon the materials, the technological history, the design, the culture, the chain of involvements -- all that we have been talking about indeed. Thus, "knowing" has to take in all that one's predecessors have done and all that one's co-workers are doing.

In the present case, whole communities were decimated by an accident of Union Carbide; under similar circumstances, say with a higher valve stress rating on the vent pipe, tank 610 might well have exploded, adding many employee casualties to the neighborhood victims. In past accidents, only several workers were injured, and in one case a worker was killed (by improperly handling phosgene). But the causes contributing to each of such accidents, whatever their consequences, are equally important. Therefore it occurs that a single inconsequential accident should have been and should be treated as and analysed as the equivalent of the accident whose consequences are enormous. The consideration of a safety violation apart from its actual or imagined consequences is an important part of safety training and philosophy.

Also, the human regard for others and the regard for others' safety are distinct factors, only moderately correlated. Often a mechanistic and highly disciplined person can better be charged with the safety of others than can a generally affectionate character. Indians, regardless of their religiousness, tend to be more solicitous and devoted to others than Northern Europeans, but they do not so readily connect their human sympathies with a determined and organized effort to ensure the safety of others -- or of themselves, for that matter.

Yet even if this statement were untrue, the distinction between safety and regard is worth preserving in personnel testing and selection. Safety-proneness must be a criterion in the recruitment and assignment of personnel, evaluated along with aptitudes, skill, and experience. There is no indication that the Union Carbide (India) collectivity was *behaviorally* aware that safety was important and that safety was measurable and insurable; to the contrary.

Union Carbide's record of safety in India has not been of the best, at least not by absolute standards of what might be readily possible: a major fire in 1978, a death by phosgene exposure of a maintenance worker in 1981; a phosgene poisoning of 23 workers by a mechanical seal breakage in 1982; burn injuries to three electricians in the control room in 1983; several burn injuries in an incident of the same year; and minor accidents of some number. Many gas leaks, no doubt implying poisonous inhalations, have been experienced.

The Indian authorities operate less actively and with far less means than American agencies. Still, the U.S. Environmental protection Agency is without adequate means generally to supervise the handling of toxic products and wastes in the U.S.A., and hitherto, state and local legislation and regulations have been often absent, under-funded, corruptly administered, or futile in preventing evasion. Considering the enormous chemical output and operations in the U.S. -- with perhaps 6000 plants handling dangerous chemicals -- the industry's largely creditable record -- not speaking now of toxic discharges and wastes but only of large-scale sudden disasters -- must be owing to the industry's self-regulation and the relatively sophisticated basic cultural comprehension of harmful chemicals. It is ironic that within the U.S. Union Carbide has been described by an officer of the Natural Resources Defence Council as having a high reputation for safety. There is a tinge of isolationism here; U.S. environmental groups have been busy with trying to "clean up their own act" and unaware that the U.S. is part of world system in safety, in the economics of production, and in the struggle to prevent wars. They are not fully politicized.

It is also notable that Union Carbide's plant at Institute, West Virginia, has reported a number of leak episodes, 28 of which were at a level requiring disclosure to the Environmental Protection Agency, 33 of which went undisclosed until in EPA investigation made further inquiries following the Bhopal tragedy. These incidents were said to be undamaging to the environment. It is once again noteworthy that Union carbide has just been fined by the EPA for not disclosing research findings related to MIC.

An American Union Carbide inspector, who visited Bhopal in 1982, recently commented on inadequacies there of training in disaster scenarios -- "what if..." exercises. Design features set conditions of disaster; small things precipitated the great disaster; ordinary procedures of maintenance and production and public information campaigns before the accident would have helped: to wit, that a company accident might harm the surrounding population, that one should, upon hearing a siren or otherwise being alerted, and upon sensing a gas irritant, cover one's face with a soaking rag and walk against the wind. That the people referred to the toxic substances made in the plant by the words "crop-medicine:" *kheti ki dawai* is nothing for a public relations staff to be proud of. Yet the same public innocence and good will could have been used to create between the factory and the community a semi-voluntary disaster-force, its personnel drawn partly from one and partly from the other, that could drill, be manifest on holiday occasions, propagandize the community, and be a source of fascination to people who are naturally curious and cooperative. The plant had a security force, which could have been made responsible for organizing, exercising and deploying a Disaster Force of one hundred men and women. Under the grim circumstances, perhaps forty would have rallied in time to be effective, if the alarm had been sounded anytime up to the massive release of gas and even for an hour afterwards. With distinctive helmets, masks, loudspeakers, and carts of cloth rags and water they would have been recognized, heard and followed. Annual costs would have been a mere $20,000 (Rs.2.5 lakhs), which even with no accidents at all could be rationalized as a worthwhile community relations expense. Fifty thousand persons might have been spared injury or death, at the worst. At best, with the alarm sounded at the recognition of an uncontrolled reaction, as many as half of all deaths and injuries might have been averted by using wet cloths and turning mass flight laterally west and east from the path of the cloud. Once the killer cloud was released there was little that the company personnel could do.

We should deal now with the background of safety failure, the larger causes. Before doing so, it may be well to note a coincidence that is infrequently realized to be a contributory factor in accidents. The accidents happened over a change of shifts at 10:45 P.M. Also the disaster occurred when the plant was practically shut down, the in-between stage of production. In warfare, a most dangerous time occurs when troops in the line are replaced by fresh troops: it is then that accidents happen, the enemy strikes, communications break down, functions are not properly transferred, intelligence is lost, confusion and panic are more likely. The analogy with the MIC plant at Bhopal on this evening is significant.

To fill out the spotty record of reported accidents, one can use a true site inspection, which is essentially a set of imagined accident scenarios. An on-site safety survey was conducted at Bhopal by a team of three Union Carbide (USA) specialists in May 1982. The survey consisted of interviews of operating personnel at all levels, and of "plant fire protection, safety and training representatives," "relevant design documents, procedures, and job safety analyses for each of the units" were reviewed. Facilities were physically inspected. One notes the absence of moral and stress tests, of emergency "game" tests, and of operations and process observations; whether because the team was incomplete or behavioral process analysis is undeveloped, the area of human behavior seems to have been absent from safety calculations. How interviews with solely Hindi-speaking personnel may have been conducted is not revealed in the text of the team report. In one place, critical mention is made of maintenance personnel issuing "permits they cannot read." In another place, high turnover of "operator (and technical personnel, to a degree)" is said to "pose a serious problem in the plant," Inadequate training of personnel who were subsequently assigned to work alone on tasks involving safety risks is also criticized.

In the course of the two and a half years following the report, which made a large number of negative findings and recommendations, all except one fault was declared corrected by June 1984. Without access to the records of those progress reports, followed by interview and investigation, one cannot take for granted the satisfactory correction of the faults mentioned; evidence exists that corrections were not made, or else the same conditions returned.

A between-the-lines reading of the report, moreover, along with the openly exposed faults and failures, allows one to surmise that there were some basically flawed conditions that the surveying team missed or were too polite, as their letter of transmission suggests, to comment upon. The large number of leaking valve occurrences was ominous, but treated as if it were a matter of going down to the nearest hardware store to replace them and of keeping them in proper working condition. Piles of junk were noted, without insisting that employees who are not disturbed about useless pipes laying around may not be keenly aware of the difference between a lethal pipe and a scrap pipe. "The housekeeping in and around the entire area (of the Sevin drums, old pipe, pools of oil on ground, etc. . ." Neglect of setting blind flanges to help the blocking of contaminants while washing out pipes was noted, with the assumption that to note the matter will bring the supervisors automatically to cure the human penchant for carelessness in this critical regard.

Another four faults are especially danger signals. One was the lack of proper identification of equipment. "Confusion as to identity could create hazardous conditions . . ." One can only presume that workers were found who could not tell one pipe from another.

The report cites "frequent pressure gauge failure" and yet there is a "high degree of reliance on manual observation and operation." Another fault is found in pipe heads that lead nowhere but are unplugged. Still further, confusion occurred when employees were interrogated as to when valves and equipment were last tested; failures to keep records on inspections and changes of equipment such as rupture discs were criticized.

On the whole, this 1982 Report was handled by the press in a thoroughly naive fashion when it was released within days of the accident. What the team said (and left unsaid) might well have given ample justification to shut down the Bhopal plant then and there, not to be reopened until practically all of the faults had been corrected, including the basic faults of an unreliable work force. The parent company never checked on the spot to insure that the deficiencies were corrected, but accepted paper assurances.

Praful Bidwai, whose articles in *The Times of India* were technically and humanistically superb, said that "it is about the filthiest and most ill maintained chemical factory I have visited anywhere. Various kinds of refuse, debris, some of it several months old, can be seen littered all over the premises. Numerous other signs of a history of long neglect and poor maintenance are visible too: corroded pipes and vessels, close to the storage tank, filth in the control room, patches of grease and foul smelling chemicals in close proximity of sensitive sections (such as that producing phosgene) of the MIC plant. . . ." A badly tended orphan, one might say, of a distant company that had not inspected for two years, and a ward of an uncaring State whose inspectors, according to an employee of many years, did nothing but stop by for the signing of forms.

The UC Inspectors' report is classified as "Business Confidential," as if the interests of the company, the public, and the State of Madhya Pradesh and City of Bhopal were unconcerned with the matter. It may be necessary to legislate measures to cause the filing of such documents with regulating authorities. That one's competitors and some Wall Street brokers may profit from a straight look into one's plant is not sufficient reason to frustrate a public interest in what is occurring within. The governments, especially when they have weak regulating units, ought not to dismiss workers' and unions' complaints but to regard them as "inspectors within the gates."

Filthy maintenance is a sign of bad morale and morale is a safety factor. Morale is to efficiency (including safety) what love is to marriage. In industry, morale is the basic positive motivation to understand one's relation to his work and to his work group and to contribute whatever one can to the effectiveness of the group's operations. Morale cannot be made up out of words; it must evolve out of the substance of men's ideals. The basic materials must be present and available.

Here I shall cite only several components of morale and ask whether they have been present in the Union Carbide setting at Bhopal:

Have the personnel of Union Carbide (India) consistently recognized the need for and importance of (not the raw fact of) a parent origin in the USA (or did they regard this tie as an anachronism or, worse, a prisoner's chains?)

Did Union Carbide at Bhopal feel appreciated at top levels in the USA and by the top management of Union Carbide (India)? What was the extent of human contacts between the two? The absence of U.S. representatives at Union Carbide (India) Board meetings was ordinary, and the only consequential visit by U.S. personnel was the aforesaid team of critical engineers. Was there anything (bonuses, thanks, compensation) to show when (as often) the Indian operation was producing higher than U.S. or average quotas of profits?

Was turnover in management personnel high? Yes. For instance, the Plant Manager was only two years on the job. Was not this bad for worker morale, too? Yes. Were Indian management and labor given to understand whether and how they were paid the equivalent of U.S. wages and other benefits? No. Were Union Carbide (India) management and personnel given to believe that they were as important as their counterparts all over the world? No. Was the Bhopal plant subjected to morale-lowering cutback? Yes. The disrepair of safety equipment shows that heavy pressure for economies was being exercised, probably under the threat, often voiced, of the closing of the plant.

Rumors of the impending sale of the plant were circulating also.

In the aftermath of accident, we find many victims asking "Was it my fate?" which converts readily to the Americanized question : "Was it my fault?" In a study of accidents among a national sample of Americans, which I once conducted for Underwriter Laboratories, a large percentage of people who had reported accidents in the family in the past year promptly blamed themselves or the victim for the accident; it appears that people are raised to take care of themselves and to find themselves at fault. To blame others is the next step, which many take, and here, among the uneducated and naïve, paranoid tendencies readily assign blame to all kinds of targets against which they harbored hostility beforehand. The scientific mind (and to lesser degree the Law of evidence in court and to a greater degree the logico-empirical philosopher) by contrast looks to the system of signals and conduct in which the accident occurs and seeks salient defects in the design of the system and the work procedures.

Cultural factors can assume major importance in setting up the parameters of a safety program. A prominent Indian (Hindu) attitude is that "Wishing will make it so," "it" being just about anything. Meanwhile a typical American attitude is, "If you don't do something to make it happen, it won't happen." (Of course, these are typical, not exclusive attitudes. In fact, the Indian attitude rendered here comes from the title of a beloved popular American song.).

If we now apply this general attitude to nature, we perceive the parallel distinctions: the Indian lives with nature, he is benevolent towards nature, and expects it to reciprocate. He lets a cow snarl traffic on a busy city street. The American wrestles with nature: he exploits nature: be tears down a natural setting and rebuilds it to his liking.

Now move from nature to science, that which deals with nature, ranging from chemical processes to human behavior. What one exploits, one mistrusts, for one expects retribution. An American then typically is more wary of nature and artifices, an Indian typically more trusting. The Indian does not anticipate adverse reactions from creatures that he has not made or harmed. The American feels that he must make and maintain things, then, whereas the

Indian may feel that things maintain themselves, and him too.

Might not some behavior reflecting this difference be present in the background and perhaps even in the instant of the Bhopal disaster? I call to mind the safety deficiencies of the Bhopal plant pointed out by the American inspectors and how long it took (2½ years) for their Indian counterparts to correct only most of these, if indeed they were truly corrected. The problems were both mechanical and human. During the same period of time, a decline was suffered in the mechanical education and aptitude of personnel attending to the scientific or natural side of the dangerous chemical conditions at the Bhopal plant.

It takes a "gauge culture" to make judgement about gauges. Persons inexperienced with gauges tend either to believe in them as infallible guides, or else to lose faith in them quickly upon any evidence of their fallibility. The "gauge-acculturated" person continuously interacts with them on the basis of a great many satisfactory or unsatisfactory encounters over a long time with gauges of all types -- in cars, refrigerators, house heating systems, at school, and at work; he worries and wonders about gauges if they are a little off or erratic, and yet tends always to take them into account, for he knows that some meaning is to be attached to them.(The same person, of course, may be ignorant of the signs that a hen is about to lay an egg, something most residents of Bhopal could tell him something about.)

At the same time as the work force was suffering reductions in educational levels, it was being reduced. One unofficial account gives the following information:

MIC Unit, Bhopal

(Personnel manning each shift)

Rank Original Number - Time of Accident

Superintendent 1 (1 for all of plant)

Supervisors 3- 1

Maintenance Supervisors 2 - 0 (at night)

Operators 12 - 6

The reduction could be justified economically by reduced sales: in 1982 with a capacity of 5000 tons of MIC-based pesticides, 2308 were produced and 2211 sold; in 1983, 1607 were produced, 1500 sold. This situation was growing critical; undeniably the business was losing money with the MIC unit, perhaps $100,000 per year. The reduction in force resulting therefrom (and the lower shift levels) affected morale and tended to render the MIC unit "accident-prone." Thus can "bottom-line" management mean lowering safety levels and enhanced potential for catastrophe.

However, suppose one were now to inquire more deeply of American culture, at the least U.S. "big business culture," whether in its own way it may be critically handicapping. The Third World and Communist press are forever accusing the American capitalist multinationals of preying for profits upon the poor countries; Americans are hardened to the allegations just as Indians may resent and rebuff any statements concerning the prevalence of the attitude of "let nature take its course." "What is wrong with cutting back on new machines, on labor, on training when profits go down? What is wrong with dumping a losing operation? What is wrong with driving a hard bargain and holding onto it relentlessly, and forcing use of one's own patents? *Caveat emptor and extremis.* It's all part of American business culture.

We begin with the remote giant corporation that pays little attention to its pesticide plant at Bhopal but expects some considerable annual tribute of dividends. It treats the operation as a set of accounting digits, and puts in charge of it westernized Indians who adopt both the technical engineering ethic that divorces the human from the machine system and the "bottom line" ethic that erases scrupulously (or should one say unscrupulously?) any intervening figures (corresponding to real human operations) contributing to lower net profits. Big business in America is run largely by anthropological automatons who are trained to look nowadays as if they "care about people" and who are helped to play this role by extensive sales, advertising, and public relations coaching. Reviewing the causes of the Bhopal disaster once more, one perceives in it the failures and weaknesses equally of Indian and American culture. The giant's remote management system and "bottom line" psychology are as critical as are outmoded and inadequate designs and equipment, the lack of training, high turnover, the poor morale and apathy, and the insistent employment of materials and processes whose unlimited risks came to be discovered with the disaster. The marriage of Indian and American culture at Bhopal was not of the best sort.

Considerations such as these may have standing in industrial psychology and cultural anthropology, but will have small place in court, where evidence of a more durable and tightly connected kind is preferred. However, in future thinking about transnational industry and commerce, it would be well to remember that transnational also means transcultural. Who can say whether, in the arguments of the seventies over setting up small containers as against large tanks for MIC storage at Bhopal, the Indians were instinctively aware of what might be better for the Indian setting as contrasted with the West Virginia site?

For that matter, who knows whether the younger generation of American workers in West Virginia has not changed, and whether there has therefore occurred an increase in "real" risk with the large tanks there; risk probabilities may be assigned to machinery and safety devices, but if human attitudes towards nature and the environment are changing, machine and safety engineering must change with them. One can only wonder, regarding that infinitely larger element of risk docked in hundreds of nuclear weapon systems, how a changing human factor is being accommodated, or even if it is being attended to from the deeper anthropological and psychological viewpoint. Secrecy is total. And if it now appears to be urgent to reconsider human and mechanical safety design throughout the world's industries in the direction of adjusting to human differences and changing capacities, how much more important is the same kind of problem in the area of the mass-killing, world-destroying armaments and war industries and operations?

Addendum:

As this book was in press, the Union Carbide "Bhopal Methyl Isocyanate Incident Investigation Team Report" of March 20, 1985 was issued. I add this note first to say what the Report is and then to say what the report is not.

The Report is a shot fired from the slit trench into which the leaders of the huge multinational Corporation have jumped. It is the only public action of Union Carbide since the first week of the Disaster. It pretends to extreme scientificity; it is narrowly technical; it can be called *scientoid*, because it aims obviously at locating all blame in India and reducing the field of industrial safety and international operations and the immense tragedy to an "incident" of the day's work.

Even so the Report, by its very issue, assumes responsibility for the disaster; one does not do *pro bono publico* technical reports and excuse oneself for something that is not one's business. Further, the Chairman, in a press release associated with the Report takes cognizance, perhaps inadvertently, of the existence of a network of "overseas locations" and of the need of bringing compliance reviews of Union Carbide facilities that handle hazardous material. . ." Union Carbide is not about to disconnect itself from all of its overseas affiliates unless and until they explode like Bhopal.

The general conclusion of the report, as we here and most others have said, is that a large volume of water entered Tank 610 containing MIC, causing an exponential reaction and explosive venting of MIC gas. It says that the "incident" happened "accidentally or deliberately;" "deliberately" -- a possibly horrendous implication -- is not defined in the glossary that consumes four pages. Indications of high amounts of chloroform in Tank 610 are present, but neither this, nor the various corrosive elements discoverable in a sample of the Tank's residues, would have altered significantly the disastrous water-MIC reaction.

The team was composed of "seven engineering and scientific specialists. . . was asserted by many other UCIL and UCC technical personnel," and, it is said, labored for nearly three months and performed "more than 500 laboratory experiments." No credit is given to the many communication specialists, legal experts, and policy-managers who also labored on the Report. Nor did the seven scientists sign their names to the Report.

The Report honestly states that "the amount of material forced out of the tank cannot be determined exactly. . . the sequence of reactions required to duplicate the composition of the core sample cannot be determined exactly."[nor was the core a valid sample]. It admits to being based largely upon hearsay evidence, but unfortunately does not let itself reach out for the broader sweep of evidence that published materials provide.

The pages of formulas for chemical reaction which are included are largely needless and could have been summarized in a prose paragraph.

The technical format and explication consists almost entirely of tests suggested by manuals of what *should* be done (but was not done) and what *might* have happened. In many places, the Report recites rules that are supposed to be followed, as if they were actually followed, lending the reading an effect of delusory thought.

The sense of psychological unreality is further enhanced by what amounts to a second-rate introductory lecture, complete with diagram, on "How to do group-thinking about an upsetting problem." This is not wrong; it is only concocted and weirdly placed.

Many critical scientific and technical questions have been overlooked or avoided. For instance, it is highly pertinent to the disaster to know more about the type of valve that failed: who designed the valves, who manufactured them, where they were obtained, how many times were complaints made to the manufacturers, did anyone connected with either the American or Indian company have anything to do with their purchase, and what were the reasons why they often failed and were not immediately replaced or repaired?

In the text of this book, I have discussed all of the features of cultural disparities, plant and instrument design, manufacturing, processing work procedures, and management that might in general and in particular have prevented the disaster. Union Carbide Corporation's Report should have addressed itself to each and every one of these, showing how it worked and how it failed.

From the failure of the Report and the corporation to address these matters, one must surmise either that the company does not have access to top-flight engineering and managerial talents, or that the format of the report was craftily drawn to engage only the narrowest of issues.

So withal, the mountain has labored and given forth a mouse. There is talk from Union Carbide of reports in the future, such as one on the medical aspects of MIC. The World does not need further Union Carbide Reports. The Corporation has shown amply that it will talk more usefully on the witness stand, in courts of law, under oath.

Lacking a world law and jurisdiction governing the behavior of multinational corporations, the Indian and American courts are both going to suffer some frustration. For example, if now the authors of the Technical Report are called before the Court, who should appear? An American Court might wish the whole writing group to show up : scientists, public relations expects, policy-making executives, lawyers -- perhaps a score of persons, or perhaps only one person who knows all the answers, perhaps the Chairman.

If the Indian Courts call all these persons, will they come? Probably not. But, if not, how can Union Carbide (USA) be questioned, in this regard or any other regard? The case would have to be heard in India without the authors of this and other critical documentation being questioned.

On March 26, the State of Madhya Pradesh Commission of Inquiry finally convened. The presiding Justice asked Union Carbide of India lawyers why they had not yet produced a report on the incident. There was no answer. Obviously Union Carbide (India) had been awaiting the report from the parent company. But, seeing that the Report from the parent company was narrowly construing the vast problem -- forcing a camel through the eye of a needle -- and was casting blame upon Union Carbide (India), the Indian company was not pleased either. After all, the Chairman of Union Carbide (USA) had just said in his news conference that "compliance with safety features is a local issue. . . That plant should not have been operating without procedures being followed." The *Hitavada* newspaper reports: "Union Carbide Counsel thereupon pointed out the UCC (USA) was different from Union carbide India and its report based on surmises and conjectures had no value. They submitted that enquiry was based on small quantities of samples from the MIC tank No. 610 which they were allowed to take sometime after the disaster."

So here now, the U.S. company is speaking for the Indian company, which however refuses to accept such representations, because it is being blamed for the accident. The head of the Commission, Justice N.K. Singh, thereupon ordered Union Carbide (India) to file a report of its own within three weeks. There may evolve some significant difference between the two companies allocating blame for the disaster. It is all grist for the mill. Still, it is already time to call off this ping-pong game between the two companies against justice. The two companies are part and parcel of the same legal complex and directed, for better or worse, by the parent company's managers. The responsibility and accountability for the Bhopal disaster must be shouldered by them jointly, and all the lesser derelictions can come trailing afterwards.

"The factory that vented the cloud"

" We had three children "

"Leading the Blind"

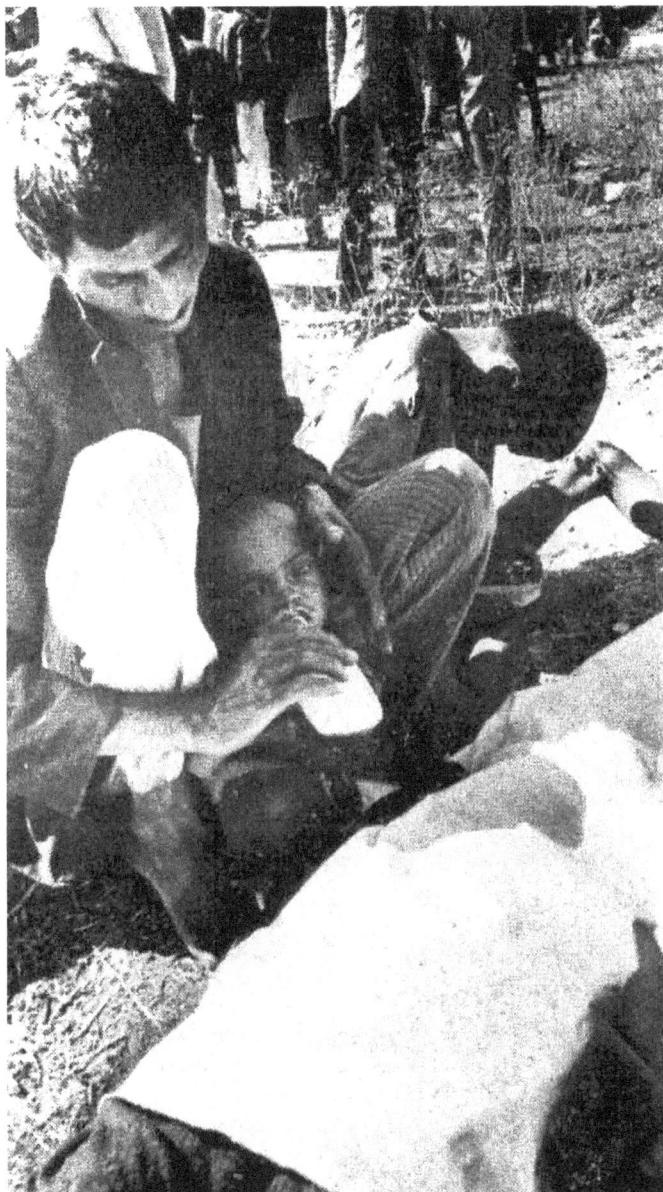

"The loss of a mother"

"Row upon row they lay"

"MIC knows no Holy Cow."

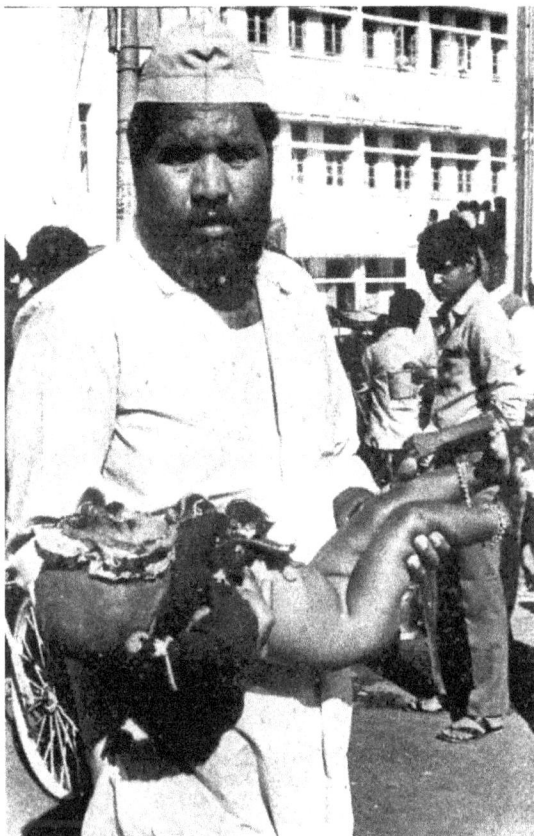

"Running for help somewhere"

"Funeral pyres burned for days"

The photographs on these pages were taken December 4-5, 1984 and are the work of Mr Mukesh Parpiani, Staff Photographer of **The Daily**, published in Bombay.

CHAPTER VII

Damages and Compensation

The cloud over Bhopal forces everyone to consider a question as old as history and philosophy: what is a person worth? The issue is very much alive in multinational circles in the world, where philosophers are rare, and will become ever more important. Ordinarily, within a given jurisdiction where one party harms another, people accept a traditional and evolving standard as to what is required to make an injured person whole again. The answer finally given to this question of making a person whole in the Bhopal situation will forever judge the judges. And judge the parties. And judge the two countries, India and the United States. And judge the state of contemporary civilization. I will say here that not only can the victims be made whole but they can even achieve a potential beyond what the past could offer them.

We must praise the leadership of Union Carbide. USA and India, for not once, in their anxiety, fright and guilt, making invidious distinctions between Indians and Americans. Such invidious distinctions may be implied from their actions of the past and present, but until now they have not resorted to degrading contrasts of relative worth such as must have quickly occurred to millions of persons attentive to the disaster in the USA, India, and the world. Probably, like other enlightened decent men and women of our age, they abhor the colonialist and racist attitudes practiced by or submitted to by our forefathers.

When a person dwells in one country and is damaged by the citizen of another country, is the damage to be weighed on the scales of the injured or the injurer? Or on both scales with a division of the difference? In the absence of principle, each party seeks to maximize his own interest, using every instrument he can command -- opinions, police, courts, law, connections, political and financial dependents and allies. Still, even in such cases, the advocacy of principle occurs, and with the addition of all elements and parties whose interests are not identical with the first parties, the determination and advocacy of principle may even become paramount in importance.

To what damages ought the hoped-for, agreed-upon principles apply? Who has been damaged in what regard? First of all, there are deaths by gassing and the complications and incidents thereof, to the number which we estimate and assume here at 3,000, and which will ultimately be arrived at by testimony of the surviving. In the absence of data, we shall assume that the dead averaged four dependents or close kin who survived. Probably 90% of all Indians work for "gain" from childhood to death. These must be provided for. A second category of persons damaged is the injured. By this time, one must appreciate that death and injury by MIC poisoning is especially painful and anguishing; dying is almost always prolonged to minutes, hours, and days; the illness, with the same harsh symptoms and suffering, is usually more prolonged and may, it now appears, extend over years of time and until death.

The injured should properly be divided into the disabled and the affected. We can assume that 10,000 persons suffer from half-disability plus a high risk of premature death. Disability occurs in several forms: in lung injury, vision impairment, muscular weakness and neurological disorders, with others, such as heart and liver problems, emerging. Some 20,000 persons can be assumed to have incurred a one-fourth disability plus a high risk of premature death.

Lesser injury, with perhaps total recovery in some cases but also perhaps lifelong impairment and progressive debilitation in others, were suffered by some 180,000 persons.

The registering, recording, diagnosing, and care of these victims and the adjustment of the survivors of the dead to a new mode of life is an immense job, beyond the present capabilities of all the voluntary and public agencies of Bhopal, Madhya Pradesh and the Union Government put together. In the above estimates, the total number of persons damaged and to be provided for in their own estimate may numbers 225,000.

We recognize that often more than one deceased comes from the same family, and therefore the calculations of compensation would in many cases double or multiply further. We must reckon on the same occurrence in regard to compensation for the injured. Should compensation be correspondingly lessened, "two for the price of one" or "three for the price of two", so to speak? Or, on the other hand, should the compensation be increased: a loss of two children to a man is more damaging than twice the loss of one? The sorrow is all the worse; the loss of potential earnings, which would have gone to an extended family, is to be multiplied proportionately, not regressively. In the balance I see no necessity to alter the formulas.

One of the nastiest arguments against awarding generous compensation to the victims comes more from bourgeois Indians than from Americans. The same argument is voiced by some who cannot wait to put money into the hands of the governments and who treat forever the people, all people, as wards of the state : "These people wouldn't know what to do with their money," it is said;' "They've never had anything. They will waste it." Americans are more tolerant of people who do not spend their money as others would.

Even many radical, anti-American Indians, who "love the poor" find the idea of a sudden enrichment of the victims of Bhopal intolerable. When pressed to explain why they feel this way, they are enveloped in embarrassment and confusion and invent various false reasons that would never apply to themselves in like circumstances. The arguments are not new to Americans, of course, who are used to poor and uneducated people becoming rich overnight, whether by inheritance, or by the discovery of oil on their land, or by winning million-dollar lotteries, or by having American Indian ancestors who were cheated of their land by governments or individuals, or by large awards in cases of accidental disablement and death as here. Instead, the Indian ought better to feel in awe of a legal system that may put wealth in the pockets of a victim instead of the state, and does not ask whether the victim is a vice-president of a bank who "knows what to do with his money."

I could well defend the right of any victim to "go to hell in his own way," given his distressing experiences, but I shall rest with the assertion that the victims will probably on the average "do as well with their money" as any random selection of a population of victims, or even of well-to-do victims, would. The reasons are indicated here and there in my report where I show that we are not dealing with fools and idiots. A poor illiterate widow who had received a small sum of money from the government told us, in the presence of a group of neighbors, that she had put the money in a bank. They agreed that she had done so and it was the right thing to do. The matter came up because she was saying: "How can money ever pay for my loss?" In any event, it is to be hoped that a proper independent trust group will act in part as an advisor to individuals and in part as a manager of rehabilitation activities on behalf of the victims.

All other losses are relatively minor in comparison with personal injuries, but in absolute terms, they are very heavy indeed. The City of Bhopal was turned into a chaos of abandonment, confusion, mourning, voluntary and emergency work, and disorganization. It is fair to say that every person in the population of about one million lost time and money. The disruption of facilities, markets, and business extended far and wide in the State of Madhya Pradesh.

It is unnecessary here to make fine distinctions, since we are establishing general parameters and principles for handling the damages and compensation. Thus differences between large and small business, types of employment and rates of earnings can be put aside momentarily. The business damages can be averaged into "business days," the figure of total loss of products, goods and services can be set at eleven "business days." There are about 300work days in the year, more than in the U.S.A. If the total annual gross product of goods and services of the City of Bhopal is figured to approximate $ 2 billion, then about 1/30 of the product will have been lost. This comes to about $ 65 million. About 200,000 workers would be eligible for compensation.

Damages of the City of Bhopal, of the hospitals and other independent public services, and of the state Government may be included under this head. In the case of hospitals, some of the damages should actually be billed as services rendered; this is an accounting formality, but should be carried properly inasmuch as payments are owing medical and hospital personnel for heavy overtime work over a period of weeks.

It may be appropriate to make contributions or awards to voluntary community agencies who performed above and beyond the call of duty, such as the People's Movement and the newspapers, mentioned earlier in these pages. Again, these are not damages, but, as payments for which revenues are sought, they can be included here.

A final item, not damages but to be reckoned with them, would be attorneys' fees. It is assumed that the cases will go to trial. It is assumed that large awards and/or settlements will be paid. If the courts allow contingency fees, the compensation of the lawyers will be heavy, exaggerated in proportion to the effort expended and large award realized. Compensation of one percent (1%) of the damages awarded should be ample and not excessive and should include expenses. This is for the courts to decide, or for those who have influence in settlement proceedings.

A final type of expense comes from executing the judgements or settlements. Once more, it is not a damage, but an administrative cost. This would cover schemes and procedures for identifying victims, medical and psychiatric examinations, follow-up research on the victims, determining net worth in many cases, statistical work, surveys and interviews. If there exist nearly a quarter of a million righteous claimants, and half that many not so righteous, procedural and medical examinations going up to a million individual cases can be expected.

As the stupendous proportions of the tragedy are thus unrolled in the detailing of damages and associated costs, the "Holocaust Syndrome" will begin to glaze the eyes of some people. "It can't have been that bad!" "The bill can never be paid!" "It's not like being hit by a car!" "One can't think in individual terms about such matters." Finally, "We are dealing with the poor, aren't we really? -- after all, they are Indians, and there are so many of them."

This line of thought must be suppressed! It is unjust, unreal, to avoid issues of holocausts, of the million dead of Verdun, of nuclear catastrophes, of mass starvation, of the Bhopal disaster. The insanity is in the psychological repression that denies the immensity of the problem and its reality. The insanity is in those who avoid responsibility, not in those who confront responsibility.

Because India has over four times the unemployment rate of the U.S.A., the Indian worker has to support four times as many people as the American worker. If his productivity in a work group averages 20% less than the American's (to be measured by the profitability of the group or another indicator), if his pay is to be equal to the American's pay (given relative costs of living), then he still must be given an extra allowance to enable him to carry his extra burden of support (which is not, as in America, carried by the social welfare system) or given to understand that his extra burden of support is a luxury, like an extra automobile in America would be, and should be reduced by population control.

The Union Carbide workers at Bhopal were well-paid by general Indian standards and even by organized industrial workers' standards. Unskilled workers received over 1000 rupees ($80 US roughly) a month, three times more than a domestic worker and 50% more than a low ranking clerical worker in government and private industry. For this reason alone, a job with UC was considered a "good job." For 1.50 rupees (about twelve cents) he could buy a lunch at the company cafeteria.

However, let us examine the case of a victim of the disaster. The typical victim was a poor worker whose recompense was so slight that it would be ignored in America as nothing at all. Yet it was significant. Children, women, the old, and the able-bodied in the Bhopal slums sought work of any kind diligently and would put to shame Americans who pride themselves on self-help and "do-it-yourself." One old man of 75 years, for instance, rolled tobacco supplied him by a dealer into tiny cigars, the "beedies." He sat in his little hut, calm and philosophical, in a setting quieted by the death of several neighbors, and rolled 1,000 beedies a day for 9.50 rupees (not an "even 10") meanwhile selling as a retail dealer for his factor a few packs of beedies on his own. Now take a little girl of 10 who watches over a goat that must wander about to find greens and garbage to eat; she also picks up twigs for the fire, cleans pots, and goes to school.

Her mother has survived, too, and is out looking for work as a hod-carrier. She will carry large stones or cement baskets on her head all day long for a few rupees; that is one type of work but she will do other types, such as guard the household, build a fire and boil water for a neighbor who has found labor on another day. Then she must sew up pieces of cloth, and shop, clean the house, bathe the baby and wash away the sewage along a sewage ditch. Can we say that the mother of the girl is worth less than the mother of an American girl, or vice versa that the child of the mother is worth less? No. probably more, if only because the interdependence is greater and the alternatives of the child fewer.

A boy may be found a mile away sitting in a tiny cobbler's shop and somehow helping the cobbler by making deliveries, picking up nails, cleaning sandals, etc. and now the man of the house, if he has survived, or if he is not away for days on a trip to his village to exchange goods, may be wandering about looking for work, or regularly sweeping out mosques and shops with his dirt-cheap broom. He may also go out to sell a few trinkets, an embroidered cloth, or to collect a load of wood in return for breaking up a large heap of it.

The people take virtually every kind of work; they do not dare to imagine criminality as employment; and they will accept almost any pay for their efforts. Furthermore, in startling contrast to Americans and Europeans, they -- these who seem to have nothing -- express a congenial mood and live in an enviable harmony of existence. If you walk among them in your fine clothes, you will not only feel safe; you will find not merely tolerance and smiles; but you will experience gratification at the compliments implied by your appearance and intimacy. You need have no love for poverty, nor be keen for snobbish slumming to understand this humanity. In the morning and evening, they bathe, showering themselves from jugs of water filled at occasional faucets on a line that the City of Bhopal has laid for them. At dark, electric bulbs light up in many houses. The streets are laid out in a grid-plan, even and well-maintained. At night, whoever are not away may sit around chewing seeds and talking, often with neighbors a few feet away, or listening to the radio. What they say is no less intelligent than the conversation among average Europeans of all classes. They are dignified, honest, integrated into a community that is real and natural, quarrelsome on occasion, devoted to children and the elderly. Although the majority of them are Moslems, they keep dogs as companions in remarkable numbers; perhaps they originated from the Hindu untouchables. These, then, are typical slum dwellers of India, perhaps 100 millions in number. These, too, are typical victims of the Bhopal disaster.

But it is time to reckon up the compensation and figure a means of paying it. We estimate the dead to have included 1500 children, 500 adults, and 1000 senior citizens. The first, say, had 40 productive years in sight, the second had 30 years, the last 10 years. The total of human years that have disappeared is 85,000.

We seek to provide an Indian with what he would need to live in a style in his society that he deems equal to that of an average American living the American way, so far as concerns diet, housing, clothing, child care, education, travel, gifts, etc. The sum of $1,500, considered a good wage in India and equal to the operator's pay and benefits at Union Carbide (Bhopal) is to be considered. Since 85,000 lost years enter our calculations to be replaced, multiplication by $1,500 gives $127,500,000 compensation for the dead. Note that penalties, anguish, suffering are not used to augment the total, but that the expectable future earnings of the dead are viewed optimistically. (Still, among the 3000 dead there might have been one future "fertilizers king" of Bombay or Singapore with lifetime earnings of $40 millions. It would not be the first time this happened, and would be more likely than the accident at Bhopal.)

For the fifty-percent disability, half the sum awarded the dead would be provided. Then ten thousand persons must receive, per year, $750 for a total of $7,500,000 per year. The number of years to be calculated might be at the same ratio as the dead, 40-30-10, with the number of young, adult, and senior guessed at 3500, 3000 and 3500. The 265,000 years at $750 give a total of $198.75 millions.

Next come the twenty-five percent disabled, 20,000 of them, who must receive $325 per year. Using the same ratio of years for the three age groups, 40-30-10, and estimating the numbers involved in the age groups at equals, one arrives at a total compensation for 553,360 defective years at $325 or $179.84 millions.

About 180,000 were affected less severely and suffered less debilitation and are less likely to suffer progressive illness. We use a figure of 10% disability for them, proceeding with the active life ratios as before (40-30-10) and with equal numbers for the age groups. Here the large number involved with a small disability results in a low personal compensation figure, for instance $150 per year for 10 years for a senior citizen, but a high total compensation for the group, $720 millions.

Turning to the business losses of Bhopal, we have already arrived at a figure of $64 millions. This could then be divided among all concerns in proportion to the number of employees. Thus a fruit vendor or taxi-driver-owner would receive $320, surely more than he would earn in two weeks of work, but at this point the formula had best be left unmodified by proportionate adjustments for the dollar-product of the business. A business employing ten persons would receive $3200.

About 2000 larger animals died. (More died later.) $100,000 can be asked for animal compensation which can be included with claims for lost valuables that may raise the amount of property loss to a million dollars (This figure includes property sold a trifle in the flight and extreme poverty that followed the disaster.) Awards to helping groups could be set at $1 million. The costs of executing judgements and/or settlements, requiring a great deal of technical and administrative services, to be discussed later, can be estimated at $20,000,000. This includes, for instance, a million medical examinations at $2.00 apiece. Finally, all of the preceding judgements must be totaled, whereupon a lawyer's fee and costs may be added, which we have set at one percent, payable even though the cases may not go to trial. It is felt that the shaping of even the awards and execution arrangements will be owing to the efforts of the attorneys in the case.

The sums work out then as follows:

Category of damages (In US $ millions)	Amount of damages
Survivors of Dead	127.50
50% Disabled	198.75
25% Disabled1	79.84
10% Disabled	720.00
Business Losses	64.00
Animal and Property Loss	1.00
Awards to Helping Groups	1.00
Costs of Executing Judgements/Settlements	20.00
Fees and costs of Attorneys	6.56
	$ 1,318.65

The total damages is about $1.3 billions then, with additional sums in lieu of damages and in payment of costs. This figure has been arrived at without exaggeration, with due regard to relative but equal human needs, and with a built-in flexibility to allow for the required precision in dealing with individual cases. Needless to say, the sum, while large because of the number of individuals involved, is modest when compared with what might be expected if the plaintiffs were residents of the United States. If such a disaster had happened in America, there would be no doubt as to the fate of Union Carbide Corporation : it would be thrown into bankruptcy.

I should point out that here I have considered the human value of life, safety, and property in India as the equal of the value of life, safety, and property everywhere. Safety I have considered as an absolute. There is no Indian standard of safety as there is an Indian life-style. At the same time, I have considered the victims and their survivors as deserving not a beggar's chance but a solid, excellent chance of restoring and fulfilling their lives. I recognize the positive functions performed by the American and Indian lawyers and the unfairness of the torrent of abuse heaped upon them. Moreover, the possibility lies in such formulas as are set forth of making many adjustments and of following additional considerations to enter without "going back to the drawing boards." Further, as I shall show, there are reasons why Union Carbide Corporation should find hope in this formula of compensation, rethink its program and future, and emerge as a leader of the new world age of chemical industry.

CHAPTER VIII

Creative Liability

Corporate officials need disaster emergency training as much as do workers, particularly in relating to the authorities. The Indian officials were caught off-guard; they were naturally shocked and before they could respond to any degree they found themselves under arrest. While the world tended to the disaster, they had to sit around for long hours and days, in the unaccustomed role of jailbirds, comfortably imprisoned, deprived of contacts with their own employees, and out of touch with their headquarters in the USA. There was no chance of throwing themselves into the vast effort to help and reorganize the community. Seven were jailed, including the Chairman from America briefly. They would ask themselves, "Am I really a criminal," and then ask it again and again, incredulously, ironically, and, of course, "What went wrong?" To which they could not well reply.

The American officers had a better chance to be
heroes. They were, after all, free and running, and had
resources. The press fell upon them, of course, ravenous
for news. Still they were able to think together and take
action. They should have called in pastors and
philosophers, but they brought in lawyers and public
relations men. The tragedy was too large for such
conventional behavior. Instead of blurting out that they
would spend every cent they could collect to help the
situation, they delved neatly into their treasury for a
dispensation of an appropriate sum of rupees with did not
look quite appropriate when translated into dollars. It
seemed as if they were already trying to settle the matter.
They were facing an irate world without a plan of defense,
or a surviving populace in Bhopal that would have burned
down the installations if it were not fearful of releasing
more MIC. Nor to this day has an adequate flow of
instruction or any constructive proposal come from the
headquarters of Union Carbide. Their apparent scheme is
to convert the mess into a technical mishap and legal case,
to fight it in the courts until doomsday and take the best
settlement offered outside the courts, meanwhile claiming
and -- who knows? -- probably sincerely, that they want
badly to settle the matter as soon as possible. Naturally
they do, but on their terms, just as the U.S. government
wants to settle the Nicaragua matter and the Soviet Union
the Afghan matter.

Union Carbide Corporation is in grave danger of
ultimate bankruptcy, arising out of its Bhopal Experience.
If the bankruptcy were rationally handled, it might not be
the worst solution, and, in fact, if the company does go
bankrupt solutions on the order of those presented here
may have a chance, if only because the handling of such a
large bankruptcy allows leeway for ingenious
reorganization of assets and meeting of obligations. Still,
the managers of Union Carbide owe it to themselves and
to the victims of Bhopal to exercise their imagination, to

come out of the trenches, to put to work in a good sense the adage that the best defense is a good offense.

What damages can Union Carbide afford to pay? We have charged it with an obligation of $1.3 billion. Union Carbide is a multinational conglomerate with current assets of $3.58 billion and current liabilities of $1.9 billion, the difference of $1.68 billion affording some indication of what funds it may have no work with in a solution. Its long-term debt is $2.3 billions, which must be serviced or might be refinanced under emergency conditions. It earned a net profit of $323 millions in 1984, after reserving $18 million as a probable loss in connection with the Bhopal tragedy. It has issued 70 million shares, 20% to institutions; of course, by its domination of many affiliates abroad, scores of thousands of shareholders overseas look hopefully to its profitability in their sector; one needs bear in mind that a decision affecting the profitability of a Union Carbide company in Germany may be made on the basis of a decision in regard to a Union Carbide company in India or America.

Union Carbide is the 3rd largest chemical concern in America, the 7th largest in the world. It is the 37th largest company overall in the USA, and as Union Carbide (India) the 25th in total sales of India. It is the 31st largest of the world's thousands of multinational companies. Its many affiliates around the world give it the international needs and relations of a medium sized nation of the world, but one would not suspect this in examining its corporate structure or leadership. It has no Department of State or Ministry of Foreign Affairs. One may note, however, that the last but one President of Union Carbide quit to become successively an ambassador and Deputy Secretary for Defense and for State. It might be argued that this acknowledges a kind of career line, but whether it is complimentary to the Department of State or to Union Carbide is questionable in view of the proven reputation of the first and the provable reputation of the second for inertness in the face of the ever more menacing complaints and aggressiveness of all three worlds towards U.S. conduct abroad.

Union Carbide (India) operates thirteen plants besides the pesticide installation at Bhopal. There it employs some 750 workers, down from over a thousand at its peak several years ago. Receipts on pesticides amounted to $14 millions in 1983, composing only 8% of the Company's total sales of $175 million. Union Carbide (India)'s equity, which gives an idea of its net worth and profitability, grew by 15 times in 30 years through 1983, from $1.7 million to $26.7 million. The company, with five divisions, produces batteries, bulbs, lamps, pesticides, films, resins, and other products. Headquarters are in Bombay, and plants are owned in seven cities.

The Press Trust of India, a new service, reported recently that in testimony before a court inquiry in Madhya Pradesh the Deputy Attorney General of the State declared that, before the disaster, Union carbide of India had decided to dismantle and sell the plant because of its unprofitability and had instituted stringent economies and staff reductions affecting safety operations.

Its Board of Directors is headed by one of the most influential and successful industrialists of India. Its managing Director (President) is a mechanical engineer, long with the company, new to his position, and paid about one-tenth of what his peers in America earn. The same ratio of one-to-ten seems to prevail for all grades of employment. Presumably, this difference is equalized in the comparable life-styles that can be afforded in the two countries. However, one may object that the "work-style" is supposed to be 100% American, that is, driven by the executive's "work ethic." And the "profit-style" may be actually reversed, so that the proportionate profits from Indian firms exceed by far the profits obtainable in the USA. What this actually means is that somehow, mentally, the Indian managerial class in a multinational situation such as this one is expected to be of triple mind -- to think

Indian for compensation, to think U.S. for production processes, and to think Indian capitalist or multinational for questions of profiting. These are not the best conditions for mental health and equanimity of soul.

What can be extracted from Union Carbide Corporation, this alchemical golden goose, without killing it? Hopeful and hateful figures are bandied about : $200 millions; $400 millions; $1 billion; $2 billions; $5, 10, 15 billions; but long before we reach the last figures we are speaking of a dead goose.

Many expert calculations will be made, based upon different premises. Here only one method of thinking about the question is suggested. One collects whatever insurance is payable. One takes from the company its Indian holdings and places the balance of the Indian Union Carbide under long-term encumbrances. One strips the parent company of its ready cash; forces it to sell practically at auction its best, not its worst, holdings, to obtain more cash; compels it to undertake new long-term obligations to trust funds set up for the Bhopal victims; and then helps it in every way possible to become the best, the most progressive, and the most profitable conglomerate multi-national operating in the traditional chemical industry that must be with us for a long time to come.

If these measures are taken, Union Carbide should be able to muster $1.3 billion to confront its Bhopal obligation -- $700 millions in the first two years (without interest), $600 millions with interest at market rates over the longer term, extending to twenty years.

How would this affect the shareholders of the parent company? If the company radically re-appraises and restructures itself for the future, it should begin to restore a dividend payment within four years, based upon a share value expected to be much lower than even at present. But shareholders and investors can behave in surprising ways. Should Union Carbide emerge like the phoenix bird from its own ashes; should it become a world leader in the organization and reform of world-wide trade and production, we should not be surprised to see it receiving a complimentary overvaluation or overconfidence in the market place. Shareholders might become patriotic and company-proud. It would snatch victory from the jaws of defeat.

More than effective internal leadership and a sound plan would be needed to redress the situation. All manner of legal obstacle will be brought to bear, whether by people who pretend to speak for victims or people who see their profit in blocking positive large solutions. For example, as I indicate below, a redistribution of securities obligations will be required, to which some stockholders might object on the basis of their immediate needs. Some few of these would undoubtedly institute legal proceedings against any conceivable plan to settle the Bhopal accounts. All the more reason that the Union Carbide plan be ambitious, generous, and progressive, for the company will need political, financial, press and public support from all quarters. The same public help can guard it against the corporate pirates who will be hoping to mobilize dissidents and take over the company.

A "Fire sale" of the Union Carbide (USA) share of Union Carbide (India) need not be required. If the company is assessed by Court appraisers to have a value, say, of $150 millions, the $75 millions belonging to Union Carbide (USA) can be turned over to the Trust Fund it is proposed to establish for the Bhopal victims. In addition, the Indian company could issue interest-bearing debentures payable over 10 tears to the amount of $25 millions, and these, too, would turned over to the Trust.

Perhaps the bulk remainder of $1.2 billions can be handled by borrowing on the market and by giving notes to be held by the Bhopal Trust, payable in decreasing amounts, with interest.

The shares of Union Carbide (USA) have been trading recently in the range of $40.00. This stock should be watered (with "holy water") by a supplementary issue of 50 million preferred shares valued at $20.00 per share, providing the Bhopal Trust with a value of $1 billion, that could be sold gradually as needed. The participation of the victims as owners of Union Carbide should let them enjoy and suffer and learn from being capitalists.

If the damages are to be settled in America, let them be settled by means congenial to American culture, so long as they do not conflict with Indian culture. Actually, the heavier interdependence and realistically greater utility of Indians to each other in extended families and beyond is congenial to semi-collective a settlement. Indians enjoy much more of a mutual support system. So one must pay to support the system as much as to support the individual. That can be borne in mind especially when one considers how the recoveries for damages should be expended and handled.

The arrangement ought to allow for both individual and system support. This may be done if the total amount of damages recovered were to be divided into three portions. The same Bhopal Trust heretofore mentioned would be assigned to administer all three parts. The Trust can be organized under Indian law under much the same conditions as in American law. The Board of Trustees would be composed of persons renowned for their independence, integrity, and skills. The American and Indian courts would designate the form of the Foundation and the Board Members in the original document of settlement. A large historical experience with this type of organization exists to be drawn upon in the USA and elsewhere. The tasks assigned the Foundation are large, numerous, complex, but no more so than those of a number of Foundations that have performed successfully in recent years.

The first portion of the damages would be paid at the earliest time to all qualified persons as soon as they can qualify by visual testimonial, and medical standards. A second portion would be held in reserve by the Foundation, with accumulating interest to be paid on an individual basis when sufficient time has elapsed to determine the long-term effects of illness upon the individuals. Any surplus or residual sum will be added to the third portion. The third portion would be used by the Foundation for system support and development whose functions would be educational and community development.

In respect to education, the Foundation would purchase from Union Carbide the present property of its Bhopal pesticides plant and from the City of Bhopal a segment of the communities devastated by the gas cloud. The structures would be maintained as a national monument to industrial safety, open to the world, with a museum for industrial safety, employing to the maximum extent as caretakers victims of the disaster.

Then, upon this same land would be erected a complete school system for Applied Science, from nursery school to postgraduate university education. Victims and their survivors would receive priority as students; all students would receive a scholarship stipend sufficient to guarantee their support, whether or not living with their families. The curriculum would be designed for education in modern life and equipped with the latest educational tools. Its aim would be to create a fully modern mentality in the young, a devotion to social work, and skills up to the limits of achievement of each student. Thus a child might enter in nursery school and complete his or her studies as a Doctor of Science and Technology, or might complete his schooling at any time along the way if his interests move in another direction, or if he cannot master the curriculum beyond a certain point. If qualified in other regards, a student changing field may attend another school with a full scholarship covering tuition and living expenses for the duration of his education.

In regard to community development, nothing less than a new city is contemplated, a city built several kilometers from the farthest limit of Bhopal by the victims and survivors, that is by the people of Bhopal to the maximum extent possible and giving the people of Bhopal a priority for transferring their residence to the new housing. The city would be built to house half a million people. The master idea employed in the new construction would be to abandon many useless traditional and modern western conceptions of housing and urban design, in favor of a design that would fulfill the needs and desires of the modern ordinary Indian with an average income. The land would be owned by the Foundation. The materials would be local, the heating solar, the electric and phone system modern and their lines buried in ground; a television cable would also be conveyed everywhere. There would be no place for private vehicles except at the town limits. The housing would be compact, with small gardens for each apartment (chickens and goats permitted). Space would be provided for public and religious construction, at the option of the surrounding residents. The city would be fireproof. There would be no elevators. And many other features that could only be supplied in a new city would be added and invented. The chief business of the new city would be modern arts and crafts, plus the building of other new cities profiting from the lessons learned here.

It would be proper to consider asking Union Carbide Corporation to set up the special corporation to build and manage the new city under control with the Foundation. Afterwards this special corporation might itself become a multinational company working not only in India but elsewhere, even in USA, to provide new cities at low costs where urban crowding has become serious and the existing cities so hopelessly out-moded and costly as to warrant their abandonment.

Settlements of this nature are practically unprecedented, and no one can be sure of its legality in American constitutional law. Tests of its legality would be sure to occur. The peculiar requirements of the case, rather than its universal importance, might persuade the

American courts to take it up in the first place and then govern its complex solution.

Yet when one compares the legal and economic complexities of the Bhopal case with the recent breakup of the giant American telecommunications monopoly into dozens of elements, each with its own territory and or functions, each with its own property and rights, and each with special local laws to observe, and altogether with a hundred million customers to accommodate and please, all of this done through the courts, one can see that the American courts have an extraordinary confidence in their ability to confront and handle problems, no matter how great, once they feel they must accept jurisdiction.

The Indian government appears inclined to negotiate a settlement with Union Carbide. If the government does so it must then file suit in the U.S. courts *parens patriae*. The Courts must approve the settlements. In this case, the American lawyers will find their assignments of attorney for the victims annulled. Various sources estimate that a settlement in the area of $300 to $400 millions may be arrived at.

Such an action on the part of the Indian government may boomerang against the government itself. The government will be called upon to administer a monstrous and thankless process, lasting for many years. Even if the Indian bureaucracy were the world's best, it would be hard pressed to pull together the necessary social work, medical, legal, accounting, *et al.* personnel and we may be sure that accusations and scandal will beset such an agency administering to the subjectively framed wants of a quarter of a million, nay, a million, people. And how will this operation be decentralized to the state of Madhya Pradesh and City of Bhopal? A great confusion and waste may be expected. How will the government ever explain administrative costs that may ultimately reach those asked by some of the American "legal vultures," who, once they have done their work, will at least fly away, while the governments will be around forever.

Is it rational for the Indian government to take on the job of taking care of up to a million individual clients? And what will poor voters everywhere think when they, mistakenly, believe that the government is treating these victims of Bhopal like pampered children while they, the others, are just as poor, often as sick, and suffer all manner of industrial and pesticidal injuries? And will the government now take upon itself the task of representing in foreign courts the case of every Indian damaged by a foreign multinational or struck by a tourist's car, for that matter? If all these risks are nevertheless to be taken, it is suggested that the government immediately rid itself of most of these troublesome tasks by turning over the whole settlement to trusts, independent of the government, such as I have described here.

Still, one more grave consideration must enter the policy of the Indian government in this case. Whatever be the total of the settlement, it will be regarded as too little. It does not matter that all "reasonable men" believe $400 millions, say, to be a just and generous settlement (and I, for one, doubt this to be so). No matter what the figure, the settlement will be considered by the political opponents of the government and by a large part of the public to be a "sell-out" to U.S. interests.

Would a billion-dollar settlement be considered a sell-out? Who knows? But now we encounter hard realities. Union Carbide will not settle on so high a figure, unless it were to take to heart the philosophy of this book, which is doubtful. So Union Carbide will only settle for a figure too low for the Indian government to avoid risk of a political disaster in accepting it.

Actually, the Indian government -- city, state, and federal -- are not unassailable with regard to responsibility for the Bhopal situation. Rather than demanding compensation from the corporations, they ought to be considering their own liability, and, besides facilitating a settlement, might well be making a larger contribution as part of the settlement, taking a token part of the burden of liability off shoulders of Union Carbide.

Consider one point further: if Union Carbide survives and flourishes after a settlement with the Indian government, a great many people will believe that Union Carbide shrewdly outwitted the Indian leaders. If Union Carbide survives and flourishes under the conditions proposed in this book, Union Carbide will be received as a "reborn Christian" or, devout Hindus will say, "deservedly reincarnated to a higher level of existence."

Union Carbide may be in a situation similar to that of the Indian Government if it deals exclusively with the Government. A "bargain" settlement of under $400 millions will find the giant company accelerated on a path of long-term decline already noted by financial experts. Wherever it moves, whether in the USA or the Third World, it will be "taxed" for its record at Bhopal. Whatever the settlement, some of its financial advisers will regard it as too high : they would prefer preemptive bankruptcy anyhow. No amount of sheer public relations will promote the settlement to a stroke of genius and good luck.

Besides it is doubtful that Union Carbide can be assured that any settlement is final. Union Carbide can be charged in American and Indian courts on scores of counts. Who gives the Indian government the right under the American Constitution to deprive a man of his day in court? Who gives such a right to Union Carbide? The American lawyers are not fools. Some of them are among the best in the business of law.

Until the settlement is legally final, Union Carbide cannot expect its financial position to be any better than it is now -- precarious and uncertain. Before the legality of the settlement is fully determined, at least a year from the settlement date, much litigation will take place in attempts to disallow the agreement. The Indian government, in its majestic sovereignty, can decide to withdraw at any time from the agreement.

It will be of help in the resolution of the overall problem if the United States government were to participate as a friend of the court and as *amicus curiae* express an intention to assist the parties to move towards a just settlement. The USA already is implicated in the case in the eyes of the world; it must concern itself with the governance of multinational companies, whether they be American or foreign companies operating in the United States and elsewhere, too, in the world economy. In the event that sums of the settlement go beyond the dimensions outlined here, the difficulties experienced by Union Carbide, whether financial or politico-economic, may threaten an excellent arrangement. The U.S. government may then choose to contribute a sum to the Trust as compensation on its own account for its indirect responsibility, or lend money to or support a public or bank-loan for Union Carbide. Its help to the Chrysler Corporation, formerly in dire straits, may offer a precedent. This is no more and even less than the U.S. government is doing for debt-ridden governments around the world who are on the brink of bankruptcy. American diplomats will hardly fail to realize the improvement of their government's relations with India that would result from this friendly service.

CHAPTER IX

The World's Chemical Crisis

When Union Carbide received a license from the Indian Government on October 31, 1975 to manufacture MIC, the government was pleased because of the relief this might afford to the foreign exchange losses implicit in importing MIC.

Around then, too, pesticides were cutting down Indian grain losses; then 25% of the crop, today's losses are 15% representing 15 million tons or enough to feed over 70 million people. Ten years later, in the wake of Bhopal, the Director of the United Nations Environment Program was saying of pesticides:

"Local regulations, inspections, monitoring, maintenance, training, education, siting, cultural differences, corporate responsibility and the transfer of technology must be reviewed directly and quickly. And it must be done with broad cooperation between governments and industry."

Large corporations operating across national boundaries do so by the consent of the nations within whose geographical limits they do their work. Their morality, their working ethics, are generally no worse and often better then those of corporations who work solely within the boundaries of the nation. Further their morality is usually no worse than that of the governments with which they deal.

The foreign corporations are licensed by the nation to operate, and the license usually betokens that they produce or bring in something that is especially desired and not adequately forthcoming from domestic corporations. For this they are usually given the privilege of taking money out of the country, this being usually their investments (or costs) plus their profits from the investments or sales. In this business, the bargains made between nations and foreign companies are sometimes better for one than the other.

If a nation already harbored corporations with the capital, resources, and skills of the foreign company, they would not let it come in, or the company would come in only on equal or worse terms than those governing domestic companies.

Nearly always, the admission of a foreign company implies or entails advantages to a nation other than those immediately obtainable in the form of production. New kinds of capital, the domestic economy, take root, and hopefully will flourish, whether directly in the field of operations or generally in the community, in years to come.

When conditions change, and what once worked to its advantage becomes onerous, a nation may have good reason and legal means for withdrawing a foreign company's licenses or increasing its obligations. A company may also withdraw from the bargain, with penalties often attached to the withdrawal.

When a foreign company withdraws from a national economy, whether voluntarily or coerced, the nation is either benefited or harmed with regards to the precise affected sector of the economy and the more extended effects referred to above.

Not surprisingly, a transnational corporation finds its operations helped or hindered by the political, economic and financial relations of its home country with its host country. Without specific fault, it can suffer from outbursts of nationalism or socialism or pacifism or religious fundamentalism, indeed, any aspect of the comportment of some or all of the political class of the host country that is or is deemed to be incompatible with its presence. To these can be added the tricky problems of fiscal transactions and legal complications. It is a sitting duck target. It is expected to behave better than local companies while at the same time it is suspected of being arrogant or know-it-all when it tries to behave better. Like the humble tourist, but with much less mobility, the multinational feels every day the variations in the economic and political weather.

Among the principal world problems, hazardous chemicals have rapidly come into prominence. Others include the related problems of pollution of land and waters; exponential population growth; conventional war and nuclear armaments; and equal justice and human rights. The problems of famine and disease, though unspeakably prevalent, are the most susceptible to administrative action, so readily available are the means for their general elimination. The largest problems represent chemistry and chemical engineering in their historical and advanced stages, including "conventional" and nuclear explosives, and the armaments industry that deals in these.

Chemistry, it must be concluded, is inextricably bound up with every major problem pressing upon mankind, except the problem of justice and human rights, that is, the proper governance of the world. It is unfortunate that from among all the science and professions, chemists as a group stand out as the least educated in and conversant with questions of politics, while most great world problems involve chemistry. It is remarkable how little chemical knowledge was to be found at all levels and at all stages of action in the Bhopal crisis. One encounters salesmen, public relations men, police, ex-military officers, mechanical engineers, lawyers, professional politicians, journalists, professional administrators, accountants, stockbrokers, insurance agents, business promoters, agitators, and professors in law and political science; yet one only encounters chemists who are called in as experts, a role as disadvantageous as it may be narrowly prestigious. Chemists, then, are an exploited group, whose fate is caused by their self-induced blindness to the political world that they have helped greatly to create.

The viper that both poor and rich states nourish at their very bosoms is the armaments industry. This, too, is a creature of chemistry. It makes a nasty contrast with the ultimately legitimate and benign chemical industry. The armaments industry is extremely hazardous, largely multinational, riddled with corruption, enveloped in secrecy through most of its operations from conception to use, accompanied by blatant advertising whose public relations managers are the governments themselves, causing in fact an infinity of fatal accidents, capable of blowing mankind and his works and life itself off the face of the Earth.

It is well for all who are concerned about the peaceful uses of chemistry to bear this in mind. Armaments are the king of hazards, the breaker of poor backs, the exploiter of human recklessness, the pamperer of degraded officialdom, the privileged dealer in hazardous chemicals down to the last bullet.

Every discussion of every problem affecting every person and group in the world ought to begin by demanding : "Destroy the weapons !"

With this clarification of issues and priorities, our attention can return to the problems of the multinational corporation. Every corporation entering from a rich country into a poor country smacks of imperialism and colonialism. The resemblance between a foreign government and a foreign corporation taking over a position in the economy is close enough to stir up bitter memories and stimulate false sensations. The chastisement of a foreign corporation, in the same way, arouses proud memories of the expulsion of the foreigners.

Despite this, and the high risks that follow, multinationals still flourish by the many thousands, some with a single branch or affiliated corporation, others with many in many countries, some specializing in a single product, others producing a broad spectrum of goods. Also, the number of multi-nationals coming out of the Third World is increasing -- there are up to 10,000 of them, with India, South Korea, Hong-Kong, Argentina and Brazil as especially prolific sires.

A large part of the world's gross product of goods and services is an outgrowth of multinational activities and any country that tries to do without them jeopardizes its economy and lowers its technological level. Yet, in receiving multinationals, a poorer nation suffers pangs of colonialism, must endure the humiliation of being a "learner" instead of a "brahmin," undergoes a loss of traditional cultural and religious values, must make and keep serious far-reaching promises over many years, and needs to pay out hard currency that it has trouble collecting. If it is forced to get dollars by borrowing, then it falls into another set of complicated and frustrating relations with the international banks, which seem to be another form of the multinational corporation, even when these appear to be fully world-oriented such as the International Monetary Fund and World Bank.

To resort to old-fashioned nationalism and parochialism reduces its economy and technology and merely makes a poor nation poorer and weaker, while to push ahead toward superior economic and technical levels at all costs raises its level of cultural frustration and disorganization without there being any world government to guarantee the future, to say "Come this way and we shall guarantee you permanently against the risks, frustrations, and failures on the way to the future world where you will enjoy all the equality and compensation you rightfully deserve."

The Bhopal disaster cannot but discourage investors in industry and commerce dealing with hazardous chemicals and encourage companies to sell of related interests. These trends are likely to occur whether or not heavy judgements or settlement costs occur with reference to Union Carbide for there is a gathering hysteria now over the government of the air, of the soil, of the water supplies; a mankind that has never learned reliably to govern the most ancient areas of life -- violence, human relations, production, population -- is entering upon new and exotic areas of rule.

What may happen, paradoxically, is that companies, whether of the West or of the Third World, that are less well equipped will plunge into such operations and that governments, too, will go more into such business, and this may not convey necessarily more safely managed enterprises and at best will bring heavily bureaucratized and high-cost production.

This last prospect -- heavier governmental participation -- may in some cases hinder worldwide agreement upon standards if governments jealous of their sovereignty refuse to lay their operations open to international inspection and to abide faithfully by worldwide rules.

In the likely event that Union Carbide pays unprecedented damages, all of the foregoing processes will be intensified and speeded up. Thus the widespread initial belief that industry and commerce worldwide will heed the fate of Union Carbide and tighten up safety practices is likely to be proven incorrect in the course of time.

The alternatives to such trends would be a true worldwide control of hazardous industries and/or the prompt substitution of new and possibly less effective, but less noxious, means of pest control everywhere. These latter solutions, or combinations of solutions, are much to be preferred to letting the world pesticide industry drift wherever it will in the aftermath of Bhopal. They run up against the sharp sentiments of neo-nationalism, the tawdry jewel of the impoverished, the poor imitation of the extravagant irresponsibility of the great powers.

We have already written of the "culture lag" accompanying high technology: Bhopal is a city most of whose people have responded to but are not possessed of an industrial culture. We can also speak of a cultural lag in the law of high technology. Just as it is negligent of a multinational corporation to set itself up to profit in a culture by equipment and procedures the culture cannot accommodate, so also it is negligent to establish itself in a legal setting where it makes its own law or benefits from an undeveloped and inadequate law. Negligent, true: still, the normal way of doing business domestically as well as internationally is often innocently and benevolently negligent. Either you are negligent or you do not do business at all, and if you do not, others will and be praised for not being arrogant, and you are depriving the culture of your technology and pushing the government against the financial wall with your hard currency exports of high technology products that the country must have, and you are wheedled and promised this and that until you accede. Whereupon eventually you become a foreigner, a public enemy.

This process has repeated itself time after time. I see no way out of its unless and until, first, a company calls upon the skills of the human sciences to acculturate its processes as they move into other cultures, as "part of the deal," introducing a full measure of innovations and education as it goes; then, secondly, that the company be part of a world movement of multinational companies to demand their own governance by world-wide rules, to demand a code of behavior under which they can live and work, where a consonance between a community's culture and the incoming technology is a serious requirement, clearly present in the minds of the corporation leaders and the government politicians and bureaucrats. An active world assembly and secretariat of multinationals, in which all functions are deliberately represented, the hazardous industries among them, should take a company by the hand as it goes into a different community setting, saying to the government: "See here, this is more than a *deal*; it is a serious engagement in which certain conditions of finance, production, acculturation, staffing, work, and safety have to be fulfilled along with whatever special *deals* and conditions you wish to make."

The ideal within a nation should become the world reality. Let me present an illustration. Following upon the Bhopal disaster, the Indian State of Tamil Nadu held up all authorizations for all chemical projects utilizing hazardous chemicals, and asked the Chemical Manufacturers' Association to report on outstanding questions, whereupon an expert team was named and reported within a month.

The report examined the Union Carbide plant of Bhopal, where hazardous chemicals, besides Methyl isocyanate, chlorine, carbon monoxide, phosgene, and methyl amine, were handled. It criticized the inadequate safety of the MIC storage tanks and the lack of medical knowledge of how to deal with the crisis. Then it recommends a number of measures:

The establishment of an independent central authority to lay down guidelines, classifications for the design of packages, shipping containers, tankers, and so on for all kinds of chemicals, with legal authority to force compliance.

The establishment of a central industrial safety and health authority to make safety rules and conduct research.

Setting up within every State of industrial response teams for cleaning up hazardous spills and related threats.

The organization by Chemical Manufacturers' Associations of emergency guidance centers on chemicals leakage and other hazards, and to report and circulate information and accident studies.

Increases in the pay and emoluments of industrial and safety inspectors to attract superior personnel.

The prompt publication and circulation of all laws and rules governing safety, health and industrial conditions. Data on toxic materials and hazards should be stored in computer data banks for ready accessibility.

Open licensing of all imported instruments and software for safety devices and computer systems, together with reduced customs duties.

Cooperation of trade unions in installing instrumentation and automatic control devices even when they result in manpower reductions.

A *cordon sanitaire*, enforced around all factories handling hazardous materials. Government help in financing relocations is recommended.

Government aid in financing safety systems and pollution controls.

Hardly any one of these recommendations can be disputed in principle. The question of who bears their financial burden arises. So also the question of finding personnel capable of administering all of the programs. Once again every bureaucracy and every government is committed to the notion that it can handle or must in any event take on every worthy program, whereas experience bluntly contradicts this notion. Once again, we revert to the educational system and the culture in general and assert that these are highly unlikely to be adequate for the tasks thrust upon their human products.

And once again we revert to the larger world where resources of education and technical experience are hoarded or bottled up. Some say of Bhopal, "it was unfortunate that the last American expert was withdrawn several years ago." That may be true: it is likely that no professional American of the chemical industry would have stood for the safety practices exhibited on all sides. It might be also that he would have received small sympathy from the U.S. side for being a troublemaker.

Several reasons explain the disappearance of the foreigner from the foreign-owned plant and one of these reasons might be sheer Indian chauvinism. Still whereas the presence of foreigners (in American as well as Indian plants) is usually beneficial and a safety measure too, it would be more generally beneficial if every government, not only Indian, could hire and use foreigners on the inspection and safety side. Officials would be horrified at the thought and quite mistaken; they need not worry, however, for the chauvinism that denies a foreigner any authority is accompanied by such poverty and miserable working conditions for the inspectorate that the turnover rate of high quality foreigners would be absurdly high.

In a state of 443,000 square kilometers, Madhya Pradesh, the government employs a score of inspectors, each of whom is assigned about 400 factories. Given 200 working days, an inspector, to complete his quota, must average to cover two factories per day, this traveling by slow bus and train. From the bus and train stops to the installation requires additional transportation, but local cabs are often unavailable or costly, and the factory management is requested to send a car. Once in the factory, only several hours in the larger factory can be afforded. If the inspection uncovers any problem, the slow wheels of government begin turning -- discussions, warnings, reports, follow-up, correspondence (no typewriters or secretarial help or private offices and little back-up clerical staff), and if a court case is decided upon (administrative rulings are only exceptionally allowed), a long delay and further efforts are entailed of the inspector; it is expectable that upcoming inspections will be perfunctory. Telephones are difficult to use in any casual continuous way to expedite business.

Inspection implies skills at inspecting something special; a *know-how* for a *what*. The two inspectors working from the Bhopal office were mechanical engineers, not chemical engineers. Inspection requires instruments, to test gauges, etc.; the inspectors had no instruments. The Labor Department inspected devices to protect worker safety, not gaseous emissions; the inspectors did call repeatedly at the Union Carbide plant after reports of mishaps and internal leaks, urging the company as a matter of course to pursue more faithfully its own procedures.

I shall not delve into the work of the Air and pollution Control Board of Madhya Pradesh, which with a small staff is supposed to inspect and discipline 200 larger and 90,000 smaller business, and municipal waste disposal as well; yet be Board has no power to issue orders "to cease and desist" harmful practices, but must go to court for this purpose. The Board did actually measure high levels of pollution surrounding the UC Bhopal plant, but remedial action did not occur.

Such problems of regulation are common in the developed areas of the world, in most of the world. The infrastructure of regulation is lacking. The costs of elevating the infrastructure to meet even present needs, much less future expectations of industrial growth, are more than can reasonably be provided in line with all other competing budget demands. In the end, under-regulation is inevitable.

What then? Abandon ideas of introducing new industry? Give up fine ideas of proper safety and working conditions? Add to the costs of doing business whether by taxes or assessments? Raise the price of the product? None of these solutions is appealing. Perhaps the most effective means of regulation would be the organization of the country's or world's business of a given category, such as pesticides, into a corporate national body and then a world body that will constitute itself to legislate appropriate safe conduct among all firms (whether privately or publicly owned). Then let this body assess its member units for insurance according to ordinary actuarial practice. As a condition of belonging to the world association, qualifying for insurance would be necessary, and in this case, inspection and regulation of a number of practices would go along with the insurance. Let the World Association conduct its own inspections, charging the cost to its annual budget with costs of problem cases thereafter charged to the company giving trouble.

In effect, a kind of world functional government would emerge, handling a large part of the burden today carried by national governments, or by no one at all. The multinational company goes beyond countries, but cannot govern well inside a country. The governments of the countries where they go are beset by and interested in domestic problems and can govern the international economic sphere only fitfully. They should be represented in, not govern, problem areas where their competences are admittedly limited. So a large place for functional self-government is foreseen in the new world economic order.

CHAPTER X

A Dynamic Memorial for Bhopal

There are lessons for the smallest child and the worldliest politician in the carnage of gas at Bhopal: "Help each other to do our work" and "Unify or Perish." Yet all the lessons of Bhopal will not live by words. The words must become operations.

A world safety manual in all languages should be prepared, whether for personal or industrial use. The axioms of safety should be taught to the child at the same age as one learns to read and write and name the continents of the globe. They should be part of a world culture. As one learns the rules of safety, one learns much about health practices, social welfare and how to work well.

A world organization of the industry and commerce of hazardous products is necessary at this time. The association should have oversight of safety education in and out of business. It should sponsor and direct worldwide a corps of auditors, consultants and inspectors, who can enter business premises anywhere that hazardous products are handled and whose advice and reports are thereafter monitored for compliance. They should be assisted by the full assembly of satellites and computerized communications and record-keeping. Hearings should be made available where business interests dispute the findings.

Sanctions of publicity, fines, suspension, expulsion, and civil or criminal court proceedings should be granted to the World Association of Hazardous Products Business.

The World Association should sponsor and establish a mutual insurance system among hazardous products business firms. The insurance premiums can be determined as a fraction of net worth, local risk factors, and company experience.

The World Association should seek to establish a situation worldwide such that no country where insufficient or unsafe conditions for safety are found to persist in law or in fact will be able to find a multinational company that will enter it on business; further, that no company going into a country will be allowed to pursue procedures banned by the Association. Thus to evade world standards would involve both a delinquent government and a delinquent company. In this case the Association will seek to prevent a delinquent company from emerging from its state to do business elsewhere.

The World Association should conduct research with the intention of substituting in every situation possible the use of non-toxic means of controlling pests, such as the use of pest-resistant varieties of seeds, using parasites for weed and insect control, introducing natural predators (plants as well as animals), pest-evasive timing of planting and harvesting, and fostering sterile male insect populations.

The World Association should monitor new biotechnological and mechanical progress and lend support to all efforts to make the newest technologies available to all countries on an equal basis regardless of their ability to pay. It may levy an assessment on its members to pay for extending the costs of new technology to the poorest countries or subventing the entrance into the country of a chosen new technology company. The World Association of Hazardous Products Business should house and provide research facilities and advice to groups engaged in monitoring and informing the world public of the hazardous activities of the multinational armaments industry.

The World Association should, whenever its activities impinge upon other types of industry and business, convene assemblies of these, replicating its own representative structure, with the idea of facilitating mutual interests and preparing for an early institution of a World Assembly of Transnational Business, whose mission would be to extend common standards of ethics, safety, compensation and working conditions everywhere.

These are some of the ideas that grew out of discussions of the writer with citizens of three continents concerning the Bhopal tragedy. The ideas need formulation at greater length. Thereupon, it is proposed that a World Congress on Safety in Commerce and Industry be convened to elaborate the ideas and enunciate insofar as possible a practical doctrine concerning them which would be promulgated worldwide. Representation at the Congress would be individual, independent, and non-governmental, with voting according to population proportions of geo-economic regions of the world.

It would be a fitting memorial to the victims of Bhopal to convene the Congress at the City of Bhopal on the First Anniversary of the tragedy, that is, at Midnight of December 2, 1985. The first order of business of the Congress would be to memorialize those who lost their lives and those who dedicated themselves to the care of the injured. Artists of the world would be asked to contribute their work to a museum on the themes of the tragedy and on world safety, peace and industrial progress. A Museum of Safety in Commerce and Industry at Bhopal would hold and exhibit the works of art and collect the historical documentation of the events. Proceedings of the Congress would be preserved there as well.

The Congress would also review and assess the progress made toward achieving justice for the Bhopal victims.

As its mottos, the Congress may adopt these:

"A Civilization qualifies as worthy to the degree that its poor enjoy a decent subsistence, a modern education, and equal justice."

"Safety is the whole World's Business."

APPENDIX

When Union Carbide received a license from the Indian Government on October 31, 1975 to manufacture MIC, the government was pleased because of the relief this might afford to the foreign exchange losses implicit in importing MIC.

Around then, too, pesticides were cutting down Indian grain losses; then 25% of the crop, today's losses are 15% representing 15 million tons or enough to feed over 70 million people. Ten years later, in the wake of Bhopal, the Director of the United Nations Environment Program was saying of pesticides:

"Local regulations, inspections, monitoring, maintenance, training, education, siting, cultural differences, corporate responsibility and the transfer of technology must be reviewed directly and quickly. And it must be done with broad cooperation between governments and industry."

Large corporations operating across national boundaries do so by the consent of the nations within whose geographical limits they do their work. Their morality, their working ethics, are generally no worse and often better than those of corporations who work solely within the boundaries of the nation. Further their morality is usually no worse than that of the governments with which they deal.

The foreign corporations are licensed by the nation to operate, and the license usually betokens that they produce or bring in something that is especially desired and not adequately forthcoming from domestic corporations. For this they are usually given the privilege of taking money out of the country, this being usually their investments (or costs) plus their profits from the investments or sales. In this business, the bargains made between nations and foreign companies are sometimes better for one than the other.

If a nation already harbored corporations with the capital, resources, and skills of the foreign company, they would not let it come in, or the company would come in only on equal or worse terms than those governing domestic companies.

Nearly always, the admission of a foreign company implies or entails advantages to a nation other than those immediately obtainable in the form of production. New kinds of capital, the domestic economy, take root, and hopefully will flourish, whether directly in the field of operations or generally in the community, in years to come.

When conditions change, and what once worked to its advantage becomes onerous, a nation may have good reason and legal means for withdrawing a foreign company's licenses or increasing its obligations. A company may also withdraw from the bargain, with penalties often attached to the withdrawal.

When a foreign company withdraws from a national economy, whether voluntarily or coerced, the nation is either benefited or harmed with regards to the precise affected sector of the economy and the more extended effects referred to above.

Not surprisingly, a transnational corporation finds its operations helped or hindered by the political, economic and financial relations of its home country with its host country. Without specific fault, it can suffer from outbursts of nationalism or socialism or pacifism or religious fundamentalism, indeed, any aspect of the comportment of some or all of the political class of the host country that is or is deemed to be incompatible with its presence. To these can be added the tricky problems of fiscal transactions and legal complications. It is a sitting duck target. It is expected to behave better than local companies while at the same time it is suspected of being arrogant or know-it-all when it tries to behave better. Like the humble tourist, but with much less mobility, the multinational feels every day the variations in the economic and political weather.

Among the principal world problems, hazardous chemicals have rapidly come into prominence. Others include the related problems of pollution of land and waters; exponential population growth; conventional war and nuclear armaments; and equal justice and human rights. The problems of famine and disease, though unspeakably prevalent, are the most susceptible to administrative action, so readily available are the means for their general elimination. The largest problems represent chemistry and chemical engineering in their historical and advanced stages, including "conventional" and nuclear explosives, and the armaments industry that deals in these.

Chemistry, it must be concluded, is inextricably bound up with every major problem pressing upon mankind, except the problem of justice and human rights, that is, the proper governance of the world. It is unfortunate that from among all the science and professions, chemists as a group stand out as the least educated in and conversant with questions of politics, while most great world problems involve chemistry. It is remarkable how little chemical knowledge was to be found at all levels and at all stages of action in the Bhopal crisis. One encounters salesmen, public relations men, police, ex-military officers, mechanical engineers, lawyers, professional politicians, journalists, professional administrators, accountants, stockbrokers, insurance agents, business promoters, agitators, and professors in law and political science; yet one only encounters chemists who are called in as experts, a role as disadvantageous as it may be narrowly prestigious. Chemists, then, are an exploited group, whose fate is caused by their self-induced blindness to the political world that they have helped greatly to create.

The viper that both poor and rich states nourish at their very bosoms is the armaments industry. This, too, is a creature of chemistry. It makes a nasty contrast with the ultimately legitimate and benign chemical industry. The armaments industry is extremely hazardous, largely multinational, riddled with corruption, enveloped in secrecy through most of its operations from conception to use, accompanied by blatant advertising whose public relations managers are the governments themselves, causing in fact an infinity of fatal accidents, capable of blowing mankind and his works and life itself off the face of the Earth.

It is well for all who are concerned about the peaceful uses of chemistry to bear this in mind. Armaments are the king of hazards, the breaker of poor backs, the exploiter of human recklessness, the pamperer of degraded officialdom, the privileged dealer in hazardous chemicals down to the last bullet.

Every discussion of every problem affecting every person and group in the world ought to begin by demanding: "Destroy the weapons!"

With this clarification of issues and priorities, our attention can return to the problems of the multinational corporation. Every corporation entering from a rich country into a poor country smacks of imperialism and colonialism. The resemblance between a foreign government and a foreign corporation taking over a position in the economy is close enough to stir up bitter memories and stimulate false sensations. The chastisement of a foreign corporation, in the same way, arouses proud memories of the expulsion of the foreigners.

Despite this, and the high risks that follow, multinationals still flourish by the many thousands, some with a single branch or affiliated corporation, others with many in many countries, some specializing in a single product, others producing a broad spectrum of goods. Also, the number of multi-nationals coming out of the Third World is increasing -- there are up to 10,000 of them, with India, South Korea, Hong-Kong, Argentina and Brazil as especially prolific sires.

A large part of the world's gross product of goods and services is an outgrowth of multinational activities and any country that tries to do without them jeopardizes its economy and lowers its technological level. Yet, in receiving multinationals, a poorer nation suffers pangs of colonialism, must endure the humiliation of being a "learner" instead of a "brahmin," undergoes a loss of traditional cultural and religious values, must make and keep serious far-reaching promises over many years, and needs to pay out hard currency that it has trouble collecting. If it is forced to get dollars by borrowing, then it falls into another set of complicated and frustrating relations with the international banks, which seem to be another form of the multinational corporation, even when these appear to be fully world-oriented such as the International Monetary Fund and World Bank.

To resort to old-fashioned nationalism and parochialism reduces its economy and technology and merely makes a poor nation poorer and weaker, while to push ahead toward superior economic and technical levels at all costs raises its level of cultural frustration and disorganization without there being any world government to guarantee the future, to say "Come this way and we shall guarantee you permanently against the risks, frustrations, and failures on the way to the future world where you will enjoy all the equality and compensation you rightfully deserve."

The Bhopal disaster cannot but discourage investors in industry and commerce dealing with hazardous chemicals and encourage companies to sell of related interests. These trends are likely to occur whether or not heavy judgements or settlement costs occur with reference to Union Carbide for there is a gathering hysteria now over the government of the air, of the soil, of the water supplies; a mankind that has never learned reliably to govern the most ancient areas of life -- violence, human relations, production, population -- is entering upon new and exotic areas of rule.

What may happen, paradoxically, is that companies, whether of the West or of the Third World, that are less well equipped will plunge into such operations and that governments, too, will go more into such business, and this may not convey necessarily more safely managed enterprises and at best will bring heavily bureaucratized and high-cost production.

This last prospect -- heavier governmental participation -- may in some cases hinder worldwide agreement upon standards if governments jealous of their sovereignty refuse to lay their operations open to international inspection and to abide faithfully by worldwide rules.

In the likely event that Union Carbide pays unprecedented damages, all of the foregoing processes will be intensified and speeded up. Thus the widespread initial belief that industry and commerce worldwide will heed the fate of Union Carbide and tighten up safety practices is likely to be proven incorrect in the course of time.

The alternatives to such trends would be a true worldwide control of hazardous industries and/or the prompt substitution of new and possibly less effective, but less noxious, means of pest control everywhere. These latter solutions, or combinations of solutions, are much to be preferred to letting the world pesticide industry drift wherever it will in the aftermath of Bhopal. They run up against the sharp sentiments of neo-nationalism, the tawdry jewel of the impoverished, the poor imitation of the extravagant irresponsibility of the great powers.

We have already written of the "culture lag" accompanying high technology: Bhopal is a city most of whose people have responded to but are not possessed of an industrial culture. We can also speak of a cultural lag in the law of high technology. Just as it is negligent of a multinational corporation to set itself up to profit in a culture by equipment and procedures the culture cannot accommodate, so also it is negligent to establish itself in a legal setting where it makes its own law or benefits from an undeveloped and inadequate law. Negligent, true: still, the normal way of doing business domestically as well as internationally is often innocently and benevolently negligent. Either you are negligent or you do not do business at all, and if you do not, others will and be praised for not being arrogant, and you are depriving the culture of your technology and pushing the government against the financial wall with your hard currency exports of high technology products that the country must have, and you are wheedled and promised this and that until you accede. Whereupon eventually you become a foreigner, a public enemy.

This process has repeated itself time after time. I see no way out of its unless and until, first, a company calls upon the skills of the human sciences to acculturate its processes as they move into other cultures, as "part of the deal," introducing a full measure of innovations and education as it goes; then, secondly, that the company be part of a world movement of multinational companies to demand their own governance by world-wide rules, to demand a code of behavior under which they can live and work, where a consonance between a community's culture and the incoming technology is a serious requirement, clearly present in the minds of the corporation leaders and the government politicians and bureaucrats. An active world assembly and secretariat of multinationals, in which all functions are deliberately represented, the hazardous industries among them, should take a company by the hand as it goes into a different community setting, saying to the government: "See here, this is more than a *deal*; it is a serious engagement in which certain conditions of finance, production, acculturation, staffing, work, and safety have to be fulfilled along with whatever special *deals* and conditions you wish to make."

The ideal within a nation should become the world reality. Let me present an illustration. Following upon the Bhopal disaster, the Indian State of Tamil Nadu held up all authorizations for all chemical projects utilizing hazardous chemicals, and asked the Chemical Manufacturers' Association to report on outstanding questions, whereupon an expert team was named and reported within a month.

The report examined the Union Carbide plant of Bhopal, where hazardous chemicals, besides Methyl isocyanate, chlorine, carbon monoxide, phosgene, and methyl amine, were handled. It criticized the inadequate safety of the MIC storage tanks and the lack of medical knowledge of how to deal with the crisis. Then it recommends a number of measures:

The establishment of an independent central authority to lay down guidelines, classifications for the design of packages, shipping containers, tankers, and so on for all kinds of chemicals, with legal authority to force compliance.

The establishment of a central industrial safety and health authority to make safety rules and conduct research.

Setting up within every State of industrial response teams for cleaning up hazardous spills and related threats.

The organization by Chemical Manufacturers' Associations of emergency guidance centers on chemicals leakage and other hazards, and to report and circulate information and accident studies.

Increases in the pay and emoluments of industrial and safety inspectors to attract superior personnel.

The prompt publication and circulation of all laws and rules governing safety, health and industrial conditions. Data on toxic materials and hazards should be stored in computer data banks for ready accessibility.

Open licensing of all imported instruments and software for safety devices and computer systems, together with reduced customs duties.

Cooperation of trade unions in installing instrumentation and automatic control devices even when they result in manpower reductions.

A *cordon sanitaire*, enforced around all factories handling hazardous materials. Government help in financing relocations is recommended.

Government aid in financing safety systems and pollution controls.

Hardly any one of these recommendations can be disputed in principle. The question of who bears their financial burden arises. So also the question of finding personnel capable of administering all of the programs. Once again every bureaucracy and every government is committed to the notion that it can handle or must in any event take on every worthy program, whereas experience bluntly contradicts this notion. Once again, we revert to the educational system and the culture in general and assert that these are highly unlikely to be adequate for the tasks thrust upon their human products.

And once again we revert to the larger world where resources of education and technical experience are hoarded or bottled up. Some say of Bhopal, "it was unfortunate that the last American expert was withdrawn several years ago." That may be true: it is likely that no professional American of the chemical industry would have stood for the safety practices exhibited on all sides. It might be also that he would have received small sympathy from the U.S. side for being a troublemaker.

Several reasons explain the disappearance of the foreigner from the foreign-owned plant and one of these reasons might be sheer Indian chauvinism. Still whereas the presence of foreigners (in American as well as Indian plants) is usually beneficial and a safety measure too, it would be more generally beneficial if every government, not only Indian, could hire and use foreigners on the inspection and safety side. Officials would be horrified at the thought and quite mistaken; they need not worry, however, for the chauvinism that denies a foreigner any authority is accompanied by such poverty and miserable working conditions for the inspectorate that the turnover rate of high quality foreigners would be absurdly high.

In a state of 443,000 square kilometers, Madhya Pradesh, the government employs a score of inspectors, each of whom is assigned about 400 factories. Given 200 working days, an inspector, to complete his quota, must average to cover two factories per day, this traveling by slow bus and train. From the bus and train stops to the installation requires additional transportation, but local cabs are often unavailable or costly, and the factory management is requested to send a car. Once in the factory, only several hours in the larger factory can be afforded. If the inspection uncovers any problem, the slow wheels of government begin turning -- discussions, warnings, reports, follow-up, correspondence (no typewriters or secretarial help or private offices and little back-up clerical staff), and if a court case is decided upon (administrative rulings are only exceptionally allowed), a long delay and further efforts are entailed of the inspector; it is expectable that upcoming inspections will be perfunctory. Telephones are difficult to use in any casual continuous way to expedite business.

Inspection implies skills at inspecting something special; a *know-how* for a *what*. The two inspectors working from the Bhopal office were mechanical engineers, not chemical engineers. Inspection requires instruments, to test gauges, etc.; the inspectors had no instruments. The Labor Department inspected devices to protect worker safety, not gaseous emissions; the inspectors did call repeatedly at the Union Carbide plant after reports of mishaps and internal leaks, urging the company as a matter of course to pursue more faithfully its own procedures.

I shall not delve into the work of the Air and pollution Control Board of Madhya Pradesh, which with a small staff is supposed to inspect and discipline 200 larger and 90,000 smaller business, and municipal waste disposal as well; yet be Board has no power to issue orders "to cease and desist" harmful practices, but must go to court for this purpose. The Board did actually measure high levels of pollution surrounding the UC Bhopal plant, but remedial action did not occur.

Such problems of regulation are common in the developed areas of the world, in most of the world. The infrastructure of regulation is lacking. The costs of elevating the infrastructure to meet even present needs, much less future expectations of industrial growth, are more than can reasonably be provided in line with all other competing budget demands. In the end, under-regulation is inevitable.

What then? Abandon ideas of introducing new industry? Give up fine ideas of proper safety and working conditions? Add to the costs of doing business whether by taxes or assessments? Raise the price of the product? None of these solutions is appealing. Perhaps the most effective means of regulation would be the organization of the country's or world's business of a given category, such as pesticides, into a corporate national body and then a world body that will constitute itself to legislate appropriate safe conduct among all firms (whether privately or publicly owned). Then let this body assess its member units for insurance according to ordinary actuarial practice. As a condition of belonging to the world association, qualifying for insurance would be necessary, and in this case, inspection and regulation of a number of practices would go along with the insurance. Let the World Association conduct its own inspections, charging the cost to its annual budget with costs of problem cases thereafter charged to the company giving trouble.

In effect, a kind of world functional government would emerge, handling a large part of the burden today carried by national governments, or by no one at all. The multinational company goes beyond countries, but cannot govern well inside a country. The governments of the countries where they go are beset by and interested in domestic problems and can govern the international economic sphere only fitfully. They should be represented in, not govern, problem areas where their competences are admittedly limited. So a large place for functional self-government is foreseen in the new world economic order.

Aftermath and Comment (1991)

When the Author of this book returned to the United States in the late Spring of 1985, he anticipated that Third World and environmental groups would have organized a strong pressure group to persuade the Union Carbide Corporation to settle generously the claims of the people of Bhopal. This had not occurred nor would it happen. There occurred a period of the filing of suits in various court jurisdictions on behalf of the victims by dozens of lawyers, acting each in the name of different individuals. Meetings were called by concerned environmental organizations, at which plans of action were devised. These proved to be pipe dreams or ineffectual. The Author urged, based on experience and theory, as expounded in his book *Kalos, What Is To Be Done With Our World?*, that direct political action and political pressures were the only means to force a just conclusion: his pleas went unheeded.

In the end, despite the depth of the tragedy and the well-intentioned efforts, it would appear that nothing could be done to move Union Carbide. The cases were consolidated from around the country in the United States District Court of Southern New York. There the numerous lawyers were grouped and ordered to speak through chosen delegates. But meanwhile the Indian Government had decided to sue on behalf of the people of Bhopal. In the face of both groups, the Court held that India was the proper jurisdictional home; there the American private attorneys were denied judicial standing, even if operating through Indian lawyers; Indian attorneys, too, were excluded.

A long period of negotiations and court proceedings ensued, with the Government of India and Union Carbide confronting one another. All of the suspiscions foreseen in this book appeared: the Indian Government was accused of taking away the rights of the victims, of settling the case behind closed doors; of making corrupt deals with Union Carbide; and so on. Union Carbide's American officials refused to go to India to testify in the proceedings.

In October of 1991, the Indian Supreme Court upheld a settlement, which had been appealed from a lower court decision of 1989, under which Union Carbide had to pay $470 million in compensation of all claims. The sum was paid over, but was not distributed to the victims, who had been receiving minute sums and medical care now and then from the Government. Officials of Bhopal claimed that the dead from the accident numbered 4000 persons.

Union Carbide got the idea of selling its 50.9% share in Union Carbide India, Ltd., worth about $70 million, in order to build a hospital for the victims in Bhopal, but the Bhopal Court forbade any such transaction, declaring that the Company should use its own funds for the purpose; further, it seized all Indian assets of Union Carbide, for ignoring the subpoenas of the Court to send its officers to India.

The Supreme Court of India found defective an earlier court order foreclosing criminal suits against officers of the Company, and opened the door to fresh charges. Thereupon, the Bhopal Court issued an arrest warrant against the now grandiosely pensioned Chairman, Warren Anderson, on charges of "culpable homicide not amounting to murder." Chief Judicial Magistrate Gulab Sharma, because the accused refused to appear, initiated an extradition order.

The process will take a long time, and the chances of Mr. Anderson ever being haled before the bar in India are remote. Still, the Author will probably be shown to have been correct in warning him in 1985 that, were he not to settle the Bhopal case generously, constructively, and amicably, he would live to regret it.

Postface (2014)

In 2004, at the occasion of the 20th anniversary of the catastrophe, stock was taken. Amnesty International upgraded to 7,000 its estimate of the number of those who had died during the horror night and in the immediate aftermath. Some 15,000 more had died in the following years as a consequence of the severe health damages they had sustained, mostly respiratory diseases, kidney failure and liver diseases. Twenty years after the catastrophe, some 30 deaths per month in Bhopal were being considered attributable to the catastrophe. The number of the chronically ill was estimated at between 100,000 and 600,000. There were still at the time 23 so-called "gas-clinics" operating in the city to provide care for the victims.

Of the $470 million fund for the compensation of the victims brokered in 1989 between the Indian Government and Union Carbide, only a fraction had yet reached the victims: those whose health had been severely and permanently affected had received a one-time settlement of $550 each; for every dead, the surviving family had received between $1,170 and $2,200. These compensations had been calculated on the basis of an average life expectancy of no more than thirty years and an average income per household of $27 per month. These numbers showed in all their starkness the brutal efficiency of the calculations of multinational companies who settled their more dangerous and labor-intensive operations in the world's poorest countries. They also displayed the staggering differences in the worth of human lives on either side of the divide between rich and poor countries.

As a measure of comparison, the families of the victims of the terrorist attack of Lockerbie were awarded 4 million US$ per victim, almost 3,500 times more. In November 2004, the Indian government ordered the filing for claims closed and the remaining 330 million dollars (including accumulated interests) to be paid out to the 572,000 victims then estimated to be alive.

The government has never been able to provide an exact number of victims or to come up with adequate scientific methods to estimate it. *« When the accord was signed in 1989, the number of persons affected by the tragedy amounted to 10 000,* according to M. S. Sharma, a journalist at *The Tribune. In 1995, when time came to pay out the money, they had become 67,000! How is it possible that, at the time of the agreement, four years after the tragedy, the number could still not have been correctly known? »*[1]

The greatest difficulty has been in identifying the victims of MIC, for nothing distinguishes drastically a patient suffering from an inhalation of MIC from another patient with respiratory disease. A geographical criterion was therefore retained, according to Olivier Bailly.

> "A medical high commission determined which areas of the city had been on the path of the gas cloud, and 36 neighborhoods of Bhopal, out of 56, were considered to have been affected, all of them situated in the Old City. In order to get a compensation, one needed to prove two things: that one had lived at that time in one of the affected neighborhoods, and that one was in possession of a medical document stating that one was ill. But in a country where, especially among the poorest, the obtaining of documents is considered a useless hassle, administrative corruption, the victims' illiteracy and the sheer dimensions of the catastrophe quickly worked against such procedures (…) *"This system has generated a trade in fake*

---

[1] Olivier Bailly, « Bhopal, l'infinie catastrophe, » in *Le Monde Diplomatique,* December 2004, http://www.monde-diplomatique.fr/2004/12/BAILLY/11723

documents," according to N.S. Sharma. *During the night of December 3rd, 1984, I was in Jammu, in Kashmir, thousands of kilometers away from Bhopal. Yet, when I arrived in Bhopal in 1991, I was offered for 800 rupees (19 dollars) a card stating that I had resided in one of the affected neighborhoods." "As for medical proofs,* adds M. Ravi Pratap Singh, from Action Aid India, *you could get them from the doctors for 1000 rupees."*

Whether victims or not, hundreds of thousands of persons have thus been taking advantage of free care for twenty years. Yet, many people, even among the poorest, do not take advantage of, or do not trust in the care provided by the government. Medications distributed by the government are reputed to be ineffective. (…) Concurring with these perceptions, M. Sattinah Sarangi allows no indulgence to the State medical services: « *These hospitals still have not developed a treatment protocol for prescriptions adapted to multiple and complex symptoms. They are run by bureaucrats and are of mediocre quality. Moreover, the government stopped, in 1994, all research into the effects of MIC, when it would have been of essential importance to continue observing over the long terms the evolution of cancers, or of the health of children who had been exposed to the gas."*

Yet, the central government and the State of Madhya Pradesh have carried on concrete projects for the welfare of the population. In addition to free health care, food rations of wheat, rice and sugar are still being distributed, twenty years after the events; workshops for professional rehabilitation have been put up, modest financial aid has been distributed, a special department has been created to handle the problems linked to the catastrophe. The discourse of the local authorities is therefore more enthusiastic: «*The government of the State of Madhya Pradesh estimates that the rehabilitation of victims has been one of the most successful of such programs ever undertaken by any government in the history of industrial disasters,"* maintains M. Bhupal Singh, senior officer of the Bhopal Gas Tragedy Relief and Rehabilitation Department. « *On the medical plane, we have succeeded in containing the consequences of the disaster. Presently, aid is sufficient.»*

A medical officer in the same department, Dr B. S. Ohri shares his analysis. « *Illnesses due to MIC correspond to a precise episode of the catastrophe, it's a confined problem. Gradually, people have found back their health and today, the situation is comfortable, there is no longer any emergency."* With 31 hospitals and government dispensaries at disposal, with a total of 634 beds, Dr Ohri even considers that the public health care offer is too vast for the 500,000 persons still affected by the catastrophe.

There is a different perception at the Nehru Hospital, situated only a few paces from the Union Carbide site: there, long queues unwind in front of the tellers, some 4,000 patients are handled there every day, according to the estimate of one orthopedist. Still, the head physician is adamant: *« Between 1987 and 1989, 362,000 persons have been examined, and 95% of them were either not sick, or they were sick temporarily. These are the observations of the physicians. The perception of the NGOs is different, but so are their interests. »* [2]

As for the ruins of the factory, they are still standing, rusty and derelict, over a 34 hectares compound overgrown with weeds. The lease ran out in 1998. The abandoned cisterns on the site are porous and their unknown contents leak onto the ground and percolate into the soil. Open-air vats carrying the name "Sevin," the name of the main pesticide once produced in the factory, are still standing around. Several studies by the government, by NGOs, by Dow Chemical itself, have testified to the high pollution of the site and to the poisoning of the ground water. A two million dollar cleaning job is supposed to have taken place, but Greenpeace estimates that a thorough cleaning and decontamination would cost 30 million.

Meanwhile, a new generation of victims is growing up in Bhopal: either children of the catastrophe, or victims of the polluted water, poisoned by chemicals. *"The factory has poisoned the soil, not only because of the 1984 catastrophe, but because of its day to day activity over many years,"* explains

---

[2] Olivier Bailly, *idem.*

Mrs Vinuta Ghopal, from Greenpeace India. (…) heavy metals have been found in the soil: zinc, copper, lead, nickel, mercury (…). Sometimes in quantities six million times superior to their natural occurrence. They have found their way into mothers' milk, and so the curse is passed on to a new generation… Twenty thousand people are thus exposed to poisonous water. On May 4, 2004, the Indian Supreme Court decreed that they had to be immediately provided with safe water. Reservoirs of 1000 litres have been put up in the affected neighborhoods, but keeping them filled is another matter (…) At Nawab Colony, where live the poorest among the poor, *"it has been three months since the reservoirs are empty. Maybe because of the ruts in the roads, the trucks cannot make it all the way to here…"*

As matters stand, the 9.3 billion dollar takeover of Union Carbide by the chemical giant Dow Chemical in 2001 has brought about nothing less than the disappearance of any legal entity that could be taken into account in such matters. Dow Chemical considers that the payment of the $470 million agreed upon has freed the corporation from any and all other responsibilities.

A trial, at last

In June 2010, more than 25 years after the catastrophe, a trial finally took place and an Indian tribunal convicted seven former managers of the plant and of Union Carbide India, and condemned them to a two year jail sentence on probation. They included the former CEO of Union Carbide India, Keshub Mahindra, the head and cofounder of Mahindra & Mahindra Limited, the automobile manufacturing corporation headquartered in Mumbai. He and his six co-defendants were condemned to a fine of $2,100 each, and the Company of Union Carbide India Limited made to pay a fine of $10,000.

After these convictions, a Union Carbide Corporation spokesperson declared: "All the appropriate people from Union Carbide India Limited – officers and those who actually ran the plant on a daily basis – have appeared to face charges."

The fate of Warren Anderson

In 2009, in response to an appeal by a Bhopal victims' group, a New Delhi court issued a warrant for the arrest of the former head of Union Carbide, Warren Anderson. The Chief judicial magistrate ordered the federal government to press Washington for his extradition.

"In Bhopal, victims and civil rights activists who gathered outside the court cheered at the news of the order. They threw slippers at an effigy of Anderson and hit it with brooms, as they danced in the streets."[3]

We remember that Warren Anderson had flown to Bhopal shortly after the tragedy and had been taken into custody for three hours upon arrival, then released on a modest bail. After which, he had promptly left India on a private airplane, thus in effect jumping bail. He has been considered a fugitive from Indian justice for failing to appear at a court hearings in 1992 where he was named chief defendant in a culpable homicide case. Anderson had gone into retirement in 1986, two years after the disaster, and had had no part in the largely disgraceful settlement reached by the Indian Government and Union Carbide. He

---

[3] *AP*, 31 July 2009, *The Guardian*.
http://www.theguardian.com/world/2009/jul/31/warren-anderson-arrest-warrant

still remained a powerful symbolic figure for the victims in Bhopal and will no doubt remain so forever.

A first request for his extradition had been made in 2003. A senior Indian foreign ministry official said the US repeatedly turned down the request for want of "evidentiary links".[4]

Following this new request of extradition, on August 1, 2009, a television crew from CBS News was dispatched to his home in the Hamptons, on Long Island, State of New York.

> "… His wife, Lillian, answered the door Saturday at the couple's modest yellow farmhouse with a white picket fence, and silver Cadillac parked in the driveway. Her husband is 89 and in poor health, she said.
>
> "We covered everything way back when," she said. "He's been haunted for many years" by the accident.
>
> Lillian Anderson wasn't aware of the new arrest warrant and said, "It's probably some political thing." She said her husband wasn't at home.
>
> "When you get to be 87 or 85 years old you just don't remember anything. You try to put bad things out of your mind," she said.
>
> Lillian Anderson said her husband has been unfairly targeted.
>
> "Every time somebody wanted to sue the company there would be some kind of a thing that happened and they would be chasing after Warren, following him to the dump with our trash," she said.

---

[4] *The Times of India,* June 10, 2010.
http://timesofindia.indiatimes.com/india/Lack-of-evidence-held-up-Anderson-extradition-MEA/articleshow/6029166.cms?referral=PM

"This is 25 years of unfair treatment, before CEOs were paid what they're paid today."[5]

In December 2010, the Indian government announced that it was determined to seek an additional compensation of one billion dollars for the catastrophe. The new request was based on the fact that the number of victims had been underestimated when the agreement in 1989 had been reached.

The 2009 extradition request for Warren Anderson was turned down by the United States authorities.

On March 22, 2011, the Indian judiciary issued a new request for the extradition of Warren Anderson, now 90 years old.

In 2012, Wikileaks revealed that Dow Chemical had hired the intelligence research organization Stratfor, of Austin, Texas, to spy extensively on the public and personal lives of activists involved in the Bhopal disaster.

Anne-Marie de Grazia

2 May 2014

---

[5] *CBS News*, August 1, 2009.
http://www.cbsnews.com/news/wife-ex-exec-haunted-by-bhopal-gas-leak/

RECENT BOOKS BY ALFRED DE GRAZIA
at Metron Publications

40 Stases & These: What Is To Be Done With Our World?
    (with art by Licia Filingeri)

God's Fire: Moses and the Management of Exodus

The Cosmic Heretics

America's History Retold (3 volumes)

A Taste of War: Soldiering in World War Two

The Iron Age of Mars

The American State of Canaan: the Peaceful, Prosperous
    Juncture of Israel and Palestine as the 51st State of
    the Unites States of America.

www.ingramcontent.com/pod-product-compliance
Lightning Source LLC
Chambersburg PA
CBHW030013290326
41934CB00005B/325